GSM – Architecture, Protocols and Services
Third Edition

GSM – Architecture, Protocols and Services

Third Edition

Jörg Eberspächer

Technische Universität München, Germany

Hans-Jörg Vögel

BMW Group Research & Technology, Germany

Christian Bettstetter

University of Klagenfurt, Austria

Christian Hartmann

Technische Universität München, Germany

A John Wiley and Sons, Ltd, Publication

This English language edition first published 2009
© 2009 John Wiley & Sons Ltd

Originally published in the German language by B.G. Teubner GmbH as "Jörg Eberspächer/Hans-Jörg Vögel/Christian Bettstetter: GSM Global System for Mobile Communication. 3. Auflage (3rd edition)."
© B.G. Teubner GmbH, Stuttgart/Leipzig/Wisbaden 2001

Registered office
John Wiley & Sons Ltd, The Atrium, Southern Gate, Chichester, West Sussex, PO19 8SQ,
United Kingdom

For details of our global editorial offices, for customer services and for information about how to apply for permission to reuse the copyright material in this book please see our website at www.wiley.com.

Library of Congress Cataloging-in-Publication Data

Eberspaecher, Joerg.
    GSM, Global System for Mobile Communication. English
    GSM : architecture, protocols and services / Joerg Eberspaecher ... [et al.]. – 3rd ed.
        p. cm.
    Prev. ed.: GSM switching, services, and protocols, 2001.
    ISBN 978-0-470-03070-7 (cloth)
1. Global system for mobile communications. I. Eberspaecher, J. (Joerg) II. Title.
    TK5103.483.E2413 2008
    621.3845'6–dc22

                                                                    2008034404

A catalogue record for this book is available from the British Library.

ISBN 978-0-470-03070-7 (H/B)

Set in 10/12pt Times by Sunrise Setting Ltd, Torquay, UK.
Printed in Great Britain by CPI Antony Rowe, Chippenham, Wiltshire.

Contents

Preface

The GSM family (GSM, GPRS, EDGE) has become one of the most successful technical innovations in history. As of June 2008, more than 2.9 billion subscribers were using GSM, corresponding to a market share of more than 81%, and its story continues, even now, despite the introduction and development of next-generation systems such as IMT-2000 or UMTS (3G) and even systems beyond 3G, dubbed IMT-Advanced.

At the same time, wireless local area networks have substantially expanded the wireless market, sometimes drawing market share from GPRS and 3G (e.g. in public WiFi hotspots), sometimes coexisting (e.g. in UMTS home routers used as a replacement for fixed wire connections). However, these are used typically for low mobility applications. Mobile communication with all of its features and stability has become increasingly important: cellular and GSM technology, plus, of course, lately 3G, GSMs sister technology, so-to-say.

Another impressive trend has emerged since our last edition: the permanent evolution in the handheld market, producing fancy mobile phones with cameras, large memory, MP3 players, Email clients and even satellite navigation. These features enable numerous nonvoice or multimedia applications, from which, of course, only a subset is or will be successful on the market.

In this third edition, we concentrate again on the architecture, protocols and operation of the GSM network and outline and explain the innovations introduced in recent years. The main novelties in this book are the presentation of capacity enhancement methods such as sectorization, the application of adaptive antennas for Spatial Filtering for Interference Reduction (SFIR) and Space Division Multiple Access (SDMA), a detailed introduction to HSCSD and EDGE for higher data rates, and an update of the available GSM services, specifically introducing the Multimedia Messaging Service (MMS).

We are happy to have received, over the past few years, many constructive comments, and a lot of praise and encouragement. The book has obviously been successfully used by professionals (especially people beginning careers in the cellular network business) but also by students including our own who use it as a textbook enhancing their course material.

Our author team has been enlarged with the addition of Dr. Christian Hartmann, an assistant professor at Technische Universität München, who took most of the load for this edition.

We thank all of the involved staff from Wiley who convinced us to prepare this updated version of a book that will hopefully be as successful over the next few years as in the past.

Jörg Eberspächer
Hans-Jörg Vögel
Christian Bettstetter
Christian Hartmann
Munich

1

Introduction

1.1 The idea of unbounded communication

Communication everywhere, with everybody, and at any time – that was the dream and goal of researchers, engineers and users, since the advent of the first wireless communication systems. Today it feels like we have almost reached that goal. Digitalization of communication systems, enormous progress in microelectronics, computers and software technology, the invention of efficient algorithms and procedures for compression, security and processing of all kinds of signals, as well as the development of flexible communication protocols have all been important prerequisites for this progress. Today, technologies are available that enable the realization of high-performance and cost-effective communication systems for many application areas.

Using current wireless communication systems, the most popular of which is GSM (Global System for Mobile Communication), we see that we have the freedom to not only roam within a network, but also between different networks, and that we can in fact communicate (almost) everywhere (unless we are in one of the rare spots still without GSM coverage today), with (almost) everybody (unless our desired communication partner is in one of the rare spots mentioned above or chooses not to be reachable), and at (almost) any time (unless we forgot to pay our last phone bill and the operator decides to lock us out). If there is one major aspect still missing in order to make our wireless experience flawless, it is the large (albeit diminishing) gap between data rates available through wireless services and those available through wired services, such as Digital Subscriber Line (xDSL). This and the limited capability of data representation at the mobile terminal (mostly due to the limited size of mobile phones) is one of the main challenges for future developments in wireless communication.

Let us now briefly take a look at the functionalities, which enable us to move and roam so freely in GSM systems: terminal mobility and personal mobility.

In the case of terminal mobility, the subscriber is connected to the network in a wireless way – via radio- or light-waves – and can move with their terminal freely, even during a

GSM – Architecture, Protocols and Services Third Edition   J. Eberspächer, H.-J. Vögel, C. Bettstetter and C. Hartmann
© 2009 John Wiley & Sons, Ltd

communication connection. The degree of mobility depends on the type of mobile radio network. The requirements for a cordless in-house telephone are much less critical than for a mobile telephone that can be used in a car or train. If mobility is to be supported across the whole network (or country) or even beyond the network (or national) boundaries, additional switching technology and administrative functions are required, to enable the subscribers to communicate in wireless mode outside of their home areas.

Such extended network functions are also needed to realize personal mobility and universal reachability. This is understood to comprise the possibility of location-independent use of all kinds of telecommunication services, including fixed and wireless networks. The user identifies themselves (the person), e.g. by using a chip card, at the place where they are currently staying and have access to the network. There, the same communication services can be used as at home, limited only by the properties of the local network or terminal used. A worldwide unique and uniform addressing system is an important requirement for personal mobility.

In the digital mobile communication system GSM, which is the subject of this book, terminal mobility is the predominant issue. Wireless communication has become possible with GSM in any town, any country and even on any continent.

GSM technology contains the essential intelligent functions for the support of personal mobility, especially with regards to user identification and authentication, and for the localization and administration of mobile users. Here it is often overlooked that in mobile communication networks by far the largest part of the communication occurs over the fixed network part, which interconnects the radio stations (base stations). Therefore, it is no surprise that in the course of further development and evolution of the telecommunication networks, a lot of thought has been given to the convergence of fixed and mobile networks.

In the beginning, GSM was used almost exclusively for speech communication; however, the Short Message Service (SMS) soon became extremely popular with GSM users: several billion text messages are being exchanged between mobile users each month. In the mean time, additional data services have been realized, most notably the High Speed Circuit Switched Data (HSCSD) and the General Packet Radio Service (GPRS), which enable improved data rate performance by allowing for more than one GSM timeslot to be used by a terminal for a service at a time. The driving factor for new (and higher bandwidth) data services obviously is wireless access to the Internet. To this end, the Wireless Application Protocol (WAP) is also explained in this book. These additions are already working towards closing the gap between wireless and fixed networks that we discussed above.

A further step was the introduction of third-generation (3G) mobile communication networks. The 3G networks, known as the Universal Mobile Telecommunication System (UMTS) in Europe and as the International Mobile Telecommunication System 2000 (IMT-2000) worldwide, have already been introduced. However, the implementation of such 3G wireless technologies has not so far stretched much beyond busy city centers. In fact, GSM is still the major technology for providing full coverage, while 3G technology is applied to cover hot-spot areas, mainly those with very high user densities. Thanks to multi-mode terminals, which can handle both standards (GSM and UMTS), wireless network users usually do not even realize which technology they are currently using while making a call or using other wireless services. Regarding the relevance of GSM technology, it is important to note that most network providers who have implemented UMTS are using basically the same fixed backbone infrastructure architecture as used for GSM and GPRS together.

1.2   The success of GSM

> GSM is now in more countries than McDonalds.
>
> *(Mike Short, Chairman MoU Association 1995–1996)*

The relevance of the GSM standard today becomes obvious when we take a brief look at the success story of GSM so far and keeping in mind that many countries are still working towards full wireless coverage, mainly by deploying GSM. GSM was initially designed as a pan-European mobile communication network, but shortly after the successful start of the first commercial networks in Europe, GSM systems were also deployed on other continents (e.g. in Australia, Hong Kong and New Zealand). In the meantime, as of May 2008, 670 networks in 208 countries are in operation according to GSM world.

In addition to GSM networks that operate in the 900 MHz frequency band, other so-called Personal Communication Networks (PCNs) and Personal Communication Systems (PCSs) are in operation. They use frequencies around 1800 MHz, or around 1900 MHz in North America. Apart from the peculiarities that result from the different frequency range, PCNs/PCSs are full GSM networks without any restrictions, in particular with respect to services and signaling protocols. International roaming among these networks is possible based on the standardized interface between mobile equipment and the Subscriber Identity Module (SIM) card, which enables personalization of equipment operating in different frequency ranges (SIM card roaming). Now that UMTS technology has been integrated by most wireless providers into their networks, roaming not only between providers but also between different technologies is already state of the art. To this end, multi-band and multi-standard terminals have been developed and are considered commonplace today. Users of state-of-the-art terminals with a SIM card from one of the major providers in Europe can use their terminals in different frequency ranges as well as in GSM and UMTS networks, without having to configure or select anything. The terminals roam between different networks and technologies automatically.

1.3   Classification of mobile communication systems

This book deals almost exclusively with GSM; however, GSM is only one of many facets of modern mobile communication.

For the bidirectional – and hence genuine – communication systems, the simplest variant is the cordless telephone with very limited mobility (particularly the Digital Enhanced Cordless Telecommunications (DECT) standard in Europe). This technology is also employed for the expansion of digital Private Branch Exchanges (PBXs) with mobile extensions.

Local Area Networks (LANs) have also been augmented with mobility functions: Wireless LANs (WLANs) have been standardized and are now offered by several companies. WLANs offer Internet Protocol (IP)-based, wireless data communication with very high bit rates but limited mobility. WLANs have been installed, for example, in office environments and airports, as a supplement or alternative to wired LANs, but also in universities, cafes, restaurants, etc. WLAN access points, however, are also very popular in private homes as access technology. In fact, in urban areas the coverage of IEEE 802.11 type access points is impressive and could theoretically be used for roaming while using WLAN by applying

Mobile IP enhanced routing for mobility support. This, however, is hindered by the fact that each WLAN cell is typically managed by someone else, in effect making it impossible to form a large network. Another aspect is that most WLAN cells are of course encrypted and cannot therefore be used by just anyone. A little different are campus-type WLAN networks, operated by companies or universities, for instance. The IEEE 802.11 type WLAN standards are continuously being amended. The IEEE 802.11n standard for high data rates enables data rates in the 100 Mbit/s range by applying multiple antennas and using multiple in multiple out (MIMO) technology. Even though standardization is not complete for IEEE 802.11n, so-called draft-n devices are already commercially available and promise data rates close to 100 Mbit/s.

Another emerging class of wireless networks are being used for short-range communication. Bluetooth, for example, replaces cables by enabling direct wireless information exchange between electronic devices (e.g. between cellular phones, Personal Digital Assistants (PDAs), computers and peripherals). These networks are also called Body Area Networks or Personal Area Networks. Unlike the mobile technologies mentioned above, they are not based on a fixed network infrastructure (e.g. base stations). The possibility of building up such networks in a spontaneous and fast way gave them the name *ad hoc* networks. WLAN technologies also include the capability for peer-to-peer *ad hoc* communication (in addition to the classical client-to-base station transmission modus).

GSM and UMTS belong to the class of cellular networks that are used predominantly for public mass communication. These had an early success with analog systems such as the Advanced Mobile Phone System (AMPS) in America, the Nordic Mobile Telephone (NMT) in Scandinavia, or the *C-Netz* in Germany. Founded on the digital system GSM (with its variants for 900, 1800 and 1900 MHz), a market with millions of subscribers worldwide was generated, and it represents an important economic force. A strongly contributing factor to this rapid development of markets and technologies has been the deregulation of the telecommunication markets, which allowed the establishment of new network operators.

Another competing or supplementary technology is satellite communication based on Low Earth Orbiting (LEO) or Medium Earth Orbiting (MEO) satellites, which also offer global, and in the long term even broadband, communication services. Trunked radio systems – in digital form with the European standard *Trans European Trunked Radio* (TETRA) – are used for business applications such as fleet control. They offer private services that are only accessible by closed user groups.

In addition to bidirectional communication systems, there also exists a variety of unidirectional systems, where subscribers can only receive but not send data. With unidirectional message systems (paging systems) users may receive short text messages. A couple of years ago, paging systems were very popular, since they offered a cost-effective reachability with wide-area coverage. Today, the SMS in GSM has basically replaced the function of paging systems. Some billion SMS messages are being exchanged between mobile GSM users each month. Digital broadcast systems, such as Digital Audio Broadcast (DAB) and Digital Video Broadcast (DVB), are very interesting for wireless transmission of radio and television stations as well as for audio- and video-on-demand and broadband transmission of Internet pages.

GSM and its enhancements (including UMTS air interfaces), however, will remain the technological base for mobile communication for many years, and will continue to open up new application areas.

1.4 Some history and statistics of GSM

In 1982 the development of a pan-European standard for digital cellular mobile radio was started by the Groupe Spécial Mobile of the CEPT (Conférence Européenne des Administrations des Postes et des Télécommunications) (see Table 1.1). Initially, the acronym GSM was derived from the name of this group. After the founding of the European standardization institute ETSI (European Telecommunication Standards Institute), the GSM group became a technical committee of ETSI in 1989. After the rapid worldwide proliferation of GSM networks, the name has been reinterpreted as Global System for Mobile Communication.

After a series of incompatible analog networks had been introduced in parallel in Europe, e.g. Total Access Communication System (TACS) in the UK, NMT in Scandinavia and the *C-Netz* in Germany, work on the definition of a European-wide standard for digital mobile radio was started in the early 1980s. The GSM was founded, which developed a set of technical recommendations and presented them to ETSI for approval. These proposals were produced by the Special Mobile Group (SMG) in working groups called Sub Technical Committees (STCs), with the following division of tasks: service aspects (SMG 01), radio aspects (SMG 02), network aspects (SMG 03), data services (SMG 04) and network operation and maintenance (SMG 06). Further working groups were mobile station testing (SMG 07), integrated circuit card aspects (SMG 09), security (SMG 10), speech aspects (SMG 11) and system architecture (SMG 12) (ETSI, 2008). SMG 05 dealt with future networks and was responsible for the initial standardization phase of the next generation of the European mobile radio system, the UMTS. Later, SMG 05 was closed, and UMTS became an independent project and technical body of ETSI. The Third Generation Partnership Project (3GPP) has been founded in cooperation with other standardization committees worldwide. Its goal was the composition of the technical specifications for UMTS. Finally, in July 2000, ETSI announced the closure of the SMG which has been responsible for setting GSM standards for the last 18 years. Their remaining and further work has been transferred to groups inside and outside ETSI; most of the ongoing work has been handed over to the 3GPP.

After the official start of the GSM networks during the summer of 1992, the number of subscribers increased rapidly such that during the fall of 1993 already more than one million subscribers had made calls in GSM networks, more than 80% of them in Germany. On a global scale, the GSM standard also received very fast recognition, as evident from the fact that at the end of 1993 several commercial GSM networks started operating outside Europe, in Australia, Hong Kong and New Zealand. Afterwards, GSM was introduced in Brunei, Cameroon, Iran, South Africa, Syria, Thailand, USA and United Arab Emirates. Whereas the majority of the GSM networks operate in the 900 MHz band (GSM900), there are also networks operating in the 1800 MHz band (GSM1800) – PCN and Digital Communication System (DCS1800) – and in the United States in the 1900 MHz band (GSM1900) – PCS. These networks use almost completely identical technology and architecture; they differ essentially only in the radio frequencies used and the pertinent high-frequency technology, such that synergy effects can be taken advantage of, and the mobile exchanges can be constructed with standard components.

In parallel to the standardization efforts of ETSI, in 1987 the then existing prospective GSM network operators and the national administrations formed a group whose members signed a common Memorandum of Understanding (MoU). The MoU Association was supposed to form a base for allowing the transnational operation of mobile stations using

Table 1.1 Time history – milestones in the evolution of GSM.

Year	Event
1982	Groupe Spécial Mobile established by the CEPT.
1986	Reservation of the 900 MHz spectrum band for GSM agreed in the EC Telecommunications Council.
	Trials of different digital radio transmission schemes and different speech codes in several countries.
1987	Basic parameters of the GSM standard agreed in February.
1988	Completion of first set of detailed GSM specifications for infrastructure.
1989	Groupe Spéciale Mobile (transferred to an ETSI technical committee) defines the GSM standard as the internationally accepted digital cellular telephony standard.
1990	GSM adaptation work started for the DCS1800 band.
1991	First GSM call made by Radiolinja in Finland.
1992	First international roaming agreement signed between Telecom Finland and Vodafone (UK).
	First SMS sent.
1993	Telstra Australia becomes the first non-European operator.
	Worlds first DCS1800 (later GSM1800) network opened in the UK.
1994	GSM Phase 2 data/fax bearer services launched.
	GSM MoU membership surpasses 100 operators.
	GSM subscribers hit one million.
1995	117 GSM networks on air.
	The number of GSM subscribers worldwide exceeds 10 million.
	Fax, data and SMS services started, video over GSM demonstrated.
	The first North American PCS 1900 (now GSM 1900) network opened.
1996	First GSM networks in Russia and China go live.
	Number of GSM subscribers hits 50 million.
1997	First tri-band handsets launched.
1998	Number of GSM subscribers worldwide over 100 million.
1999	WAP trials begin in France and Italy.
2000	First commercial GPRS services launched.
	First GPRS handsets enter the market.
	Five billion SMS messages sent in one month.
2001	First 3GSM (W-CDMA) network goes live.
	Number of GSM subscribers exceed 500 million worldwide.
2003	First EDGE networks go live.
	Membership of GSM Association breaks through 200-country barrier.
	Over half a billion handsets produced in a year.
2008	GSM surpasses three billion customer threshold.

internationally standardized interfaces. As of April 2008, the GSM MoU has 747 members which operates 670 GSM networks in 200 countries.

1.5   Overview of the book

The remainder of this book is organized as follows. In Chapter 2, we give an introduction to radio channel characteristics and the cellular principle. The understanding of duplex and multiple access schemes serves as the basis for understanding GSM technology. We also describe some measures to increase the capacity in GSM systems, sectorization, as applied by most GSM networks already today, and Spacial Filtering for Interference Reduction (SFIR). Chapter 3 introduces the GSM system architecture and addressing. It explains the basic structure and elements of a GSM system and their interfaces as well as the identifiers of users, equipment and system areas. Next, Chapter 4 deals with the physical layer at the air interface (how are speech and data transmitted over the radio channel?). Among other things, it describes GSM modulation, multiple access, duplexing, frequency hopping, the logical channels and synchronization. Also we discuss GSM coding (source coding, speech processing and channel coding). In Chapter 5, the entire protocol architecture of GSM (payload transport and signaling) is covered. For example, communication protocols for radio resource management, mobility management, connection management at the air interface are explained as well as mechanisms for authentication and encryption. Chapter 6 describes in detail three main principles that are needed for roaming and switching: location registration and update (i.e. how does the network keep track of the user and find them when there is an incoming call?), connection establishment and termination and handover (i.e. how is a call transferred between cells?). Chapter 8 is on enhanced data services in GSM. It explains in detail GPRS which can be used for wireless Internet access. In addition this chapter includes HSCSD and Enhanced Date Rates for Global Evolution (EDGE). Chapter 7 contains the major GSM services and, finally, Chapter 9 gives a brief outlook on future mobile network developments. Appendix A covers basic GSM data services and Appendix B describes network operation and management.

2

The mobile radio channel and the cellular principle

Many measures, functions and protocols in digital mobile radio networks are based on the properties of the radio channel and its specific qualities, in contrast to information transmission through guided media. For the understanding of digital mobile radio networks it is therefore helpful to know a few related basic principles. For this reason, the most important fundamentals of the radio channel and of cellular and transmission technology are presented and briefly explained in the following. For a more detailed treatment, see, for example, Bertsekas and Gallager (1987), Lee (1989), Proakis (1995) and Steele and Hanzo (1999).

2.1 Characteristics of the mobile radio channel

The electromagnetic wave of the radio signal propagates under ideal conditions in free space in a radial-symmetric pattern. The received power P_r decreases with the square of the distance L from the transmitter. Specifically, the received power P_r can be described according to the free-space model as a function of the transmit power P_t, the distance L and the wavelength of the radio signal λ as

$$P_r = P_t \cdot g_t \cdot g_r \cdot \left(\frac{\lambda}{4\pi L}\right)^2, \qquad (2.1)$$

where g_t and g_r are the transmit and receive antenna gains, respectively. While this model is appropriate, for instance, for inter-satellite as well as for Earth-to-satellite communication, it does not capture the effects of terrestrial radio propagation, where the signal is scattered and reflected by obstacles such as buildings, mountains, vegetation, the ground and water surfaces. At the receiver, direct and – potentially many – reflected signal components are superimposed. In effect, we can describe P_r as a linear function of P_t, g_t, g_r, and an overall channel gain g_c:

$$P_r = g_c \cdot g_t \cdot g_r \cdot P_t. \qquad (2.2)$$

GSM – Architecture, Protocols and Services Third Edition   J. Eberspächer, H.-J. Vögel, C. Bettstetter and C. Hartmann
© 2009 John Wiley & Sons, Ltd

The channel gain g_c can be split into three components

$$g_c = g_d(L) \cdot g_s \cdot g_m \qquad (2.3)$$

each capturing one of the main propagation effects.

- **Distance-dependent path gain $g_d(L)$:** This part of the channel gain is usually modeled as a deterministic function of the distance L between the transmitter and the receiver, such that $g_d(L) \cdot P_t$ gives the mean received power at distance L from the transmitter (assuming $g_t = g_r = 1$). A common model for the path gain is given by

$$g_d(L) = \left(\frac{\lambda}{4\pi L} \right)^2 \left(\frac{L_0}{L} \right)^{\gamma-2} \sim L^{-\gamma}, \qquad (2.4)$$

  where L_0 is a reference distance and $\gamma \geq 2$ is the attenuation exponent, depending on the propagation environment (Rappaport, 2002). Typical values for γ are between 3 and 5. In addition to the described model, specifically for modeling and planning of GSM networks, measurement-based models are available, such as the Okumura–Hata model (Hata, 1980; Okumura, 1968) for GSM900 networks and the COST-231 Hata model (Damosso, 1999) for GSM1800 networks. Those models are parameterized by the heights of transmit and receive antennas as well as by the propagation environment (rural, sub-urban or urban).

- **Shadowing gain g_s:** Shadowing describes the effect of fluctuations of the received power around the main value, as it is caused by obstacles such as buildings and vegetation. The severeness of the shadowing effect depends on the number and properties of obstacles between the transmitter and receiver. Changes in shadowing occur in the order of meters, e.g. when a user turns around a corner during a phone call. In accordance with measurement data, the most commonly used model for shadowing is a statistical model, describing the shadowing gain g_s as a log-normal distributed random variable. Therefore, the shadowing gain in decibels, i.e. $\chi = 10 \log_{10}(g_s)$, is distributed according to a Gaussian distribution given by

$$f_\chi(\chi) = \frac{1}{\sqrt{2\pi}\sigma} \cdot e^{-\chi^2/2\sigma^2}. \qquad (2.5)$$

  The standard deviation σ defines the severeness of the shadowing and depends on the environment to be modeled. According to measurements, typical values for σ are between 5 and 10 dB (Geng and Wiesbeck, 1998).

- **Multipath fading gain g_m:** Another source of received power fluctuations around the mean value is caused by multipath propagation. In urban environments, in particular, multiple copies of the transmitted signal arrive at the receiver through different propagation paths. The superposition of many such copies of the transmitted signal, arriving at the receiver from different directions and with different delays, causes a wave field around the receiver. The received signal strength within this wave field changes severely in the order of the signal wavelength between places where destructive and constructive superposition occurs. The resulting amplitude variations

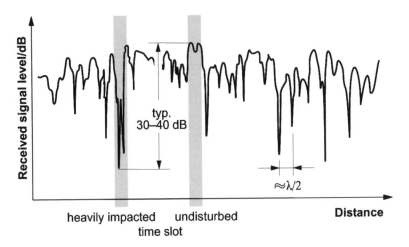

Figure 2.1 Typical signal in a channel with Rayleigh fading.

are modeled by a random variable a, such that

$$g_m = a^2. \tag{2.6}$$

The distribution of the random variable a depends on the propagation environment. If no direct line of sight between sender and receiver is present, a is assumed to be Rayleigh distributed, while an additional line of sight can be taken into consideration if a Rice distribution is applied. Figure 2.1 shows typical channel fluctuations according to Rayleigh fading for a receiver traveling through the wave field. It can be shown that if a is Rayleigh distributed, the multipath fading gain $g_m = a^2$ will be exponentially distributed (Schwartz, 2005).

The signal level observed at a specific location is determined by the phase shift of the multipath signal components. This phase shift depends on the wavelength of the signal, and thus the signal level at a fixed location is also dependent on the transmission frequency. Therefore, the fading phenomena in radio communication are also frequency specific. If the bandwidth of the mobile radio channel is small (narrowband signal), then the whole frequency band of this channel is subject to the same propagation conditions, and the mobile radio channel is considered frequency-nonselective. On the other hand, if the bandwidth of a channel is large (broadband signal), the individual frequencies suffer from different degrees of fading (Figure 2.2) in which case we speak of a frequency-selective channel (David and Benkner, 1996; Steele, 1992). Signal breaks because of frequency-selective fading along a signal path are much less frequent for a broadband signal than for a narrowband signal, because the fading holes only shift within the band and the received total signal energy remains relatively constant (Bossert, 1991).

In addition to frequency-selective fading, the different propagation times of the individual multipath components also cause time dispersion on their propagation paths. Therefore, signal distortions can occur due to interference of one symbol with its neighboring symbols ('intersymbol interference'). These distortions depend first on the spread experienced by a

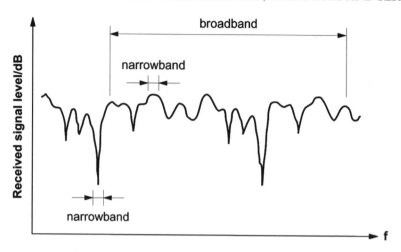

Figure 2.2 Frequency selectivity of a mobile radio channel.

pulse on the mobile channel, and second on the duration of the symbol or of the interval between symbols. Typical multipath channel delays range from 0.5 μs in urban areas to about 16 to 20 μs in hilly terrain, i.e. a transmitted pulse generates several echoes which reach the receiver with delays of up to 20 μs. In digital mobile radio systems with typical symbol durations of a few microseconds, this can lead to smearing of individual pulses over several symbol durations.

Owing to the described effects of the wireless channel, mobile information transport requires additional, often very extensive measures, which compensate for the effects of multipath propagation. First, an equalizer is required, which attempts to eliminate the signal distortions caused by intersymbol interference. The operational principle of such an equalizer for mobile radio is based on the estimation of the channel pulse response to periodically transmitted, well-known bit patterns, known as the training sequences (Bertsekas and Gallager, 1987; Watson, 1993). This allows the time dispersion of the channel and its compensation to be determined. The performance of the equalizer has a significant effect on the quality of the digital transmission. On the other hand, for efficient transmission in digital mobile radio, channel coding measures are indispensable, such as forward error correction with error-correcting codes, which allows the effective bit error ratio to be reduced to a tolerable value (about 10^{-5} to 10^{-6}). Further important measures are transmitter power control and algorithms for the compensation of signal interruptions in fading, which may be of such a short duration that a disconnection of the call would not be appropriate.

2.2 Separation of directions and duplex transmission

The most frequent form of communication is the bidirectional communication which allows simultaneous transmitting and receiving. A system capable of doing this is called full-duplex. One can also achieve full-duplex capability if sending and receiving do not occur simultaneously but switching between both phases is done so fast that it is not noticed by the user, i.e. both directions can be used quasi-simultaneously. Modern digital mobile radio systems are always full-duplex capable. Essentially, two basic duplex procedures are employed: Frequency Division Duplex (FDD) using different frequency bands in each direction, and Time Division Duplex (TDD) which periodically switches the direction of transmission.

2.2.1 Frequency Division Duplex

The frequency duplex procedure has been used already in analog mobile radio systems and is also used in digital systems. For communication between a mobile and a base station, the available frequency band is split into two partial bands, to enable simultaneous sending and receiving. One partial band is assigned for *uplink* (from mobile to base station) transmissions and the other partial band is assigned for *downlink* (from base station to mobile) transmissions.

- Uplink band: transmission band of the mobile *and* receiving band of the base station.

- Downlink band: receiving band of the mobile *and* transmission band of the base station.

To achieve good separation of both directions, the partial bands must be a sufficient frequency distance apart, i.e. the frequency pairs of a connection assigned to uplink and downlink must have this distance band between them. Usually, the same antenna is used for sending and receiving. A duplexing unit is then used for the directional separation, consisting essentially of two narrowband filters with steep flanks (Figure 2.3). These filters, however, cannot be integrated, so pure frequency duplexing is not appropriate for systems with small compact equipment (David and Benkner, 1996).

2.2.2 Time Division Duplex

Time duplexing is therefore a good alternative, especially in digital systems with time division multiple access. In this case, the transmitter and receiver operate only quasi-simultaneously at different points in time, i.e. the directional separation is achieved by switching in time between transmission and reception, and thus no duplexing unit is required. Switching occurs frequently enough that the communication appears to be over a quasi-simultaneous full-duplex connection. However, out of the periodic interval T available for the transmission of a time slot only a small part can be used, so that a time duplex system requires more than twice the bit rate of a frequency duplex system.

2.3 Multiple access

The radio channel is a communication medium shared by many subscribers in one cell. Mobile stations compete with one another for the frequency resource to transmit their information streams. Without any other measures to control simultaneous access of several users, collisions can occur (multiple access problem). Since collisions are very undesirable for a connection-oriented communication like mobile telephony, the individual subscribers/mobile stations must be assigned dedicated channels on demand. In order to divide the available physical resources of a mobile system, i.e. the frequency bands, into voice channels, special multiple access procedures are used which are presented in the following (Figure 2.4).

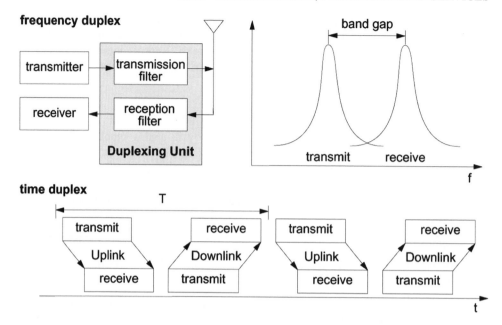

Figure 2.3 Frequency and time duplex.

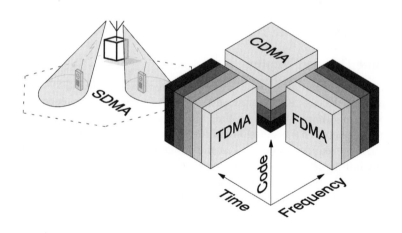

Figure 2.4 Multiple access procedures.

2.3.1   Frequency Division Multiple Access

Frequency Division Multiple Access (FDMA) is one of the most common multiple access procedures. The frequency band is divided into channels of certain bandwidth such that each conversation is carried on a different frequency (Figure 2.5). The effort in the base station to realize an FDMA system is very high. Even though the required hardware components are

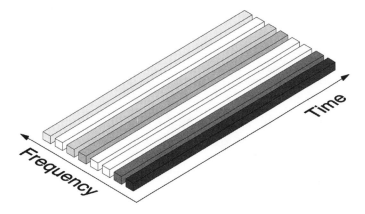

Figure 2.5 Channels of an FDMA system.

relatively simple, each channel needs its own transceiving unit. Furthermore, the tolerance requirements for the high-frequency networks and the linearity of the amplifiers in the transmitter stages of the base station are quite high, since a large number of channels need to be amplified and transmitted together (David and Benkner, 1996; Steele, 1992). One also needs a duplexing unit with filters for the transmitter and receiver units to enable full-duplex operation, which makes it hard to build small, compact mobile stations, since the required narrowband filters can hardly be realized with integrated circuits.

2.3.2   Time Division Multiple Access

Time Division Multiple Access (TDMA) is used in digital mobile radio systems. The individual mobile stations are cyclically assigned a frequency for exclusive use only for the duration of a time slot, which obviously requires frame synchronization between transmitter and receiver. Furthermore, in most cases the whole system bandwidth for a time slot is not assigned to one station, but the system frequency range is subdivided into subbands, and TDMA is used for multiple access to each subband. The subbands are known as carrier frequencies, and the mobile systems using this technique are designated as multicarrier systems (not to be confused with multicarrier modulation). GSM employs such a combination of FDMA and TDMA; it is a multicarrier TDMA system. The available frequency range is divided into frequency channels of 200 kHz bandwidth each (with guard bands between to ease filtering), with each of these frequency channels containing eight TDMA conversation channels.

Thus, the sequence of time slots assigned to a mobile station represents the physical channels of a TDMA system. In each time slot, the mobile station transmits a data burst. The period assigned to a time slot for a mobile station thus also determines the number of TDMA channels on a carrier frequency. The time slots of one period are combined into a so-called TDMA frame. Figure 2.6 shows five channels in a TDMA system with a period of four time slots and three carrier frequencies.

The TDMA signal transmitted on a carrier frequency in general requires more bandwidth than an FDMA signal; this is because with multiple time use, the gross data rate has to be

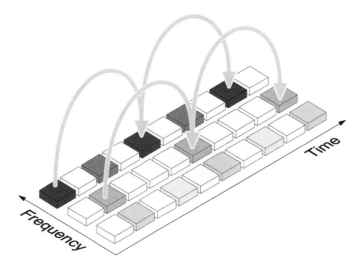

Figure 2.6 TDMA channels on multiple carrier frequencies.

correspondingly higher. For example, GSM systems employ a gross data rate (modulation data rate) of 271 kbit/s on a subband of 200 kHz, which amounts to 33.9 kbit/s for each of the eight time slots.

Narrowband systems are particularly susceptible to frequency-selective fading (Figures 2.1 and 2.2) as already mentioned, such that a single channel might be in a deep fade while switching to another channel might result in a significantly better reception. Furthermore, there are also frequency-selective co-channel interferences, which can contribute to the deterioration of the transmission quality. To this end a TDMA system offers very good opportunities to attack and drastically reduce such frequency-selective interference by introducing a frequency hopping technique. With this technique, each burst of a TDMA channel is transmitted on a different frequency (Figure 2.7).

In this technique, selective interference on one frequency at worst hits only every i_{th} time slot, if there are i frequencies available for hopping. Thus, the signal transmitted by a frequency hopping technique uses frequency diversity. Of course, the hopping sequences must be orthogonal, i.e. one must ascertain that two stations transmitting in the same time slot do not use the same frequency. Since the duration of a hopping period is long compared with the duration of a symbol, this technique is called slow frequency hopping. With fast frequency hopping, the hopping period is shorter than a time slot and is of the order of a single symbol duration or even less. This technique belongs to the family of spread spectrum techniques. As mentioned above, for TDMA, synchronization between a mobile and base station is necessary. This synchronization becomes even more complex due to the mobility of the subscribers, because they can stay at varying distances from the base station and their signals thus incur varying propagation times. First, the basic problem is determining the exact moment when to transmit. This is typically achieved by using one of the signals as a time reference, such as the signal from the base station (downlink, Figure 2.8). On receiving the TDMA frame from the base station, the mobile can synchronize and transmit a time slot

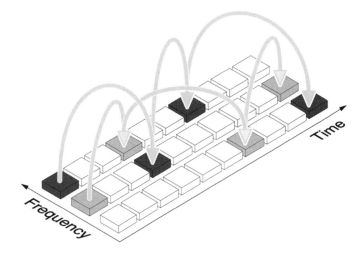

Figure 2.7 TDMA with the use of a frequency hopping technique.

synchronously with an additional time offset (e.g. three time slots in Figure 2.8). Another problem is the propagation time of the signals, ignored up to now. It also depends on the variable distance of the mobile from the base station. These propagation times are the reason why the signals that arrive on the uplink are not frame-synchronized at the base, but have variable delays. If these delays are not compensated, collisions of adjacent time slots can occur (Figure 2.8). In principle, the mobile stations must therefore advance the offset time between reception and transmission, i.e. the start of sending, so that the signals arrive in a frame-synchronous manner at the base station.

2.3.3   Code Division Multiple Access

Systems with Code Division Multiple Access (CDMA) are broadband systems, in which each subscriber uses the whole system bandwidth (similar to TDMA) for the complete duration of the connection (similar to FDMA). However, usage is not exclusive, i.e. all of the subscribers in a cell use the same frequency band simultaneously. To separate the signals, the subscribers are assigned orthogonal codes. The basis of CDMA is a band-spreading or spread spectrum technique. The signal of one subscriber is spread spectrally over a multiple of its original bandwidth. Typically, spreading factors are between 10 and 1000; they generate a broadband signal for transmission from the narrowband signal, and this is less sensitive to frequency-selective interference and disturbances. Furthermore, the spectral power density is decreased by band spreading, and communication is even possible below the noise threshold (David and Benkner, 1996).

Direct sequence CDMA

A common spread-spectrum procedure is the direct sequence technique (Figure 2.9). Here, each bit of the original data sequence is multiplied directly – before modulation – with a

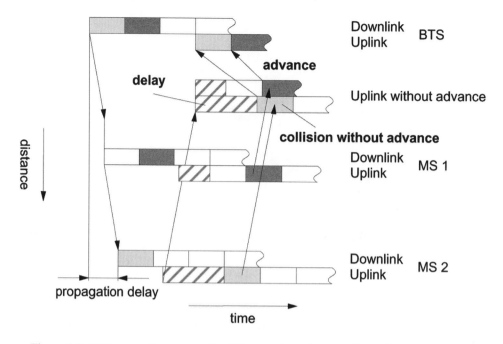

Figure 2.8 Differences in propagation delays and synchronization in TDMA systems.

spreading sequence or part of a spreading sequence to generate the spread spectrum signal. The length of the spreading sequence with respect to the duration of an original data bit defines the Spreading Factor (SF). Consequently, the transmission rate after spreading, which is called the chip rate, is larger than the original data bit rate (by the SF). In a multiple access system, the spread signals of many users are superimposed and at the receiver, the composite signal is correlated with the user specific spreading sequences in order to retrieve the original user data sequence (Figure 2.10).

Thus, if direct sequence spreading is used, the multiple access scheme is called Direct Sequence CDMA (DS-CDMA).

Frequency hopping CDMA

Another possibility for spreading the band is the use of a fast frequency hopping technique. If one changes the frequency several times during one transmitted data symbol, a similar spreading effect occurs as in the case of the direct sequence procedure. If the frequency hopping sequence is again controlled by orthogonal code sequences, another multiple access system can be realized, the Frequency Hopping CDMA (FH-CDMA).

2.3.4   Space Division Multiple Access

An essential property of the mobile radio channel is multipath propagation, which leads to frequency-selective fading phenomena. Furthermore, multipath propagation is the cause of another significant property of the mobile radio channel, the spatial fanning-out of signals.

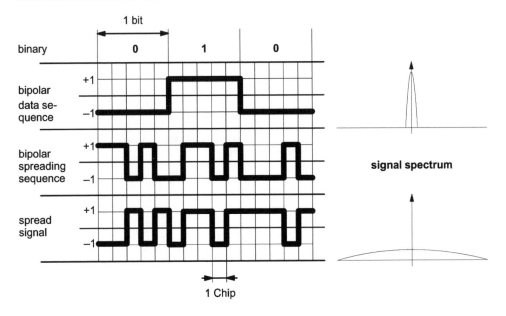

Figure 2.9 Principle of spread spectrum technique for direct sequence CDMA.

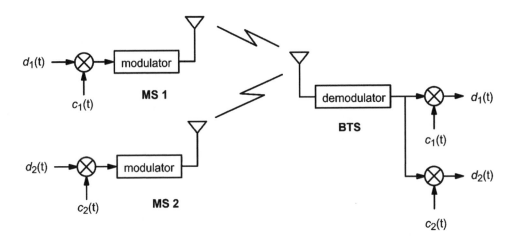

Figure 2.10 Simplified scheme of CDMA (uplink).

This causes the received signal to be a summation signal, which is not only determined by the Line of Sight (LOS) connection but also by an undetermined number of individual paths caused by refractions, infractions and reflections. In principle, the directions of incidence of these multipath components could therefore be distributed arbitrarily at the receiver. In particular, on the uplink from the mobile station to the base station, there is, however, in most

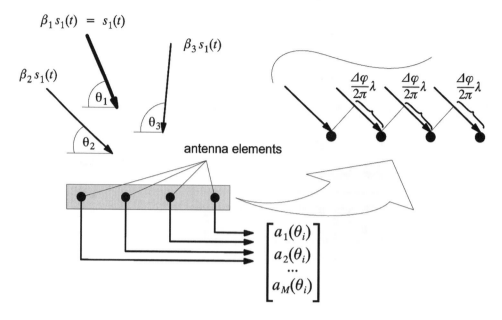

Figure 2.11 Multipath signal at an antenna array.

cases a main direction of incidence (usually LOS), about which the angles of incidence of the individual signal components are scattered in a relatively narrow range. Frequently, the essential signal portion at the receiver is distributed only over an angle of a few degrees. This is because base stations are installed wherever possible as free-standing units, and there are no interference centers in the immediate neighborhood. This directional selectivity of the mobile radio channel, which exists in spite of multipath propagation, can be exploited by using antenna arrays. These generate a directional characteristic by controlling the phases of the signals from the individual antenna elements. This allows the receiver to adjust the antenna selectively to the main direction of incidence of the received signal, and conversely to transmit selectively in one direction. This principle can be illustrated easily with a simple model (Figure 2.11).

The individual multipath components $\beta_i s_1(t)$ of a transmitted signal $s_1(t)$ propagate on different paths such that the multipath components incident at an antenna under the angle θ_i differ in amplitude and phase. If one considers an array antenna with M elements ($M = 4$ in Figure 2.11) and a wave front of a multipath component incident at angle θ_i on this array antenna, then the received signals at the antenna elements differ mainly in their phase – each shifted by $\Delta\phi$ (Figure 2.11) – and amplitude.

In this way, the response of the antenna to a signal incident at angle θ_i can be characterized by the complex response vector $\mathbf{a}(\theta_i)$ which defines amplitude gain and phase of each antenna element relative to the first antenna element ($a_1 = 1$):

$$\mathbf{a}(\theta_i) = [a_1(\theta_i)\ a_2(\theta_i)\ \cdots\ a_M(\theta_i)]^{\mathrm{T}} = [1\ a_2(\theta_i)\ \cdots\ a_M(\theta_i)]^{\mathrm{T}}, \qquad (2.7)$$

where the superscript T denotes the transpose of a vector. The N_m multipath components ($N_m = 3$ in Figure 2.11) of a signal $s_1(t)$ generate, depending on the incidence angle θ_i, a received signal vector $\mathbf{x}_1(t)$ which can be written with the respective antenna response vector and the signal of the ith multipath $\beta_i s_1(t)$ shifted in amplitude and phase against the direct path $s_1(t)$ as

$$\mathbf{x}_1(t) = \mathbf{a}(\theta_1)s_1(t) + \sum_{i=2}^{N_m} \mathbf{a}(\theta_i)s_1(t) = \mathbf{a}_1 s_1(t). \tag{2.8}$$

In this case, the vector $\mathbf{a}_1$ is also designated the spatial signature of the signal $s_1(t)$ which remains constant as long as the source of the signal does not move and the propagation conditions do not change (Xu and Li, 1994). In a multi-access situation, there are typically several (N_q) sources. This yields the following result for the total signal at the array antenna, neglecting noise and interferences

$$\mathbf{x}(t) = \sum_{j=1}^{N_q} \mathbf{a}_j s_j(t). \tag{2.9}$$

From this summation signal, the signals of the individual sources are separated by weighting the received signals of the individual antenna elements with a complex factor (weight vector $\mathbf{w}_i$), which yields

$$\mathbf{w}_i^H \mathbf{a}_j = \begin{cases} 1 & \text{if } i = j, \\ 0 & \text{if } i \neq j. \end{cases} \tag{2.10}$$

For the weighted summation signal (Xu and Li, 1994) we obtain

$$\mathbf{w}_i^H \mathbf{x}(t) = \sum_{j=1}^{N_q} \mathbf{w}_i^H \mathbf{a}_j s_j(t) = s_i(t). \tag{2.11}$$

Under ideal conditions, i.e. neglecting noise and interference, the signal $s_i(t)$ of a single source i can be separated from the summation signal of the array antenna by using an appropriate weight vector during signal processing. The determination of the optimal weight vector, however, is a nontrivial and computation-intensive task. Owing to the considerable processing effort and also because of the mechanical dimensions of the antenna field, array antennas are predominantly used in base stations. So far only the receiving direction has been considered. The corresponding principles, however, can also be used for constructing the directional characteristics of the transmitter. Assume symmetric propagation conditions in the sending and receiving directions, and assume the transmitted signals $s_i(t)$ are weighted with the same weight vector $\mathbf{w}_i$ as the received signal, before they are transmitted through the array antenna; then one obtains the following summation signal radiated by the array antenna:

$$\mathbf{y}(t) = \sum_{j=1}^{N_q} \mathbf{w}_j s_j(t) \tag{2.12}$$

and for the signal received on the ith opposite side, respectively:

$$\hat{s}_i(t) = \mathbf{a}_i^H \mathbf{y}(t) = \sum_{j=1}^{N_q} \mathbf{a}_i^H \mathbf{w}_j s_j(t) = s_i(t). \tag{2.13}$$

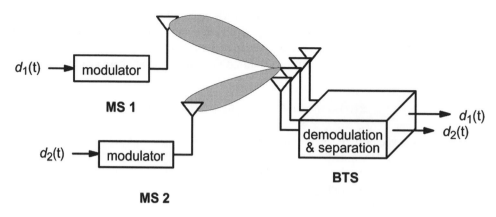

Figure 2.12 Schematic representation of spatial multiple access (uplink).

Thus, by using array antennas, one can separate the simultaneously received signals of spatially separated subscribers by exploiting the directional selectivity of the mobile radio channel. Owing to the use of intelligent signal processing and corresponding control algorithms, such systems are also known as systems with intelligent antennas. The directional characteristics of the array antenna can be controlled adaptively such that a signal is only received or transmitted in exactly the spatial segment where a certain mobile station is currently staying. On the one hand, one can thus reduce co-channel interference in other cells, and on the other hand, the sensitivity against interference can be reduced in the current cell. Furthermore, because of the spatial separation, physical channels in a cell can be reused, and the lobes of the antenna diagram can adaptively follow the movement of mobile stations. In this case, yet another multiple access technique (Figure 2.12) is defined and known as Space Division Multiple Access (SDMA). SDMA systems are currently the subject of intensive research. The SDMA technique can be combined with each of the other multiple access techniques (FDMA, TDMA, CDMA). This enables intracellular spatial channel reuse, which again increases the network capacity (Hartmann and Vögel, 1999). This is especially attractive for existing networks which can use an intelligent implementation of SDMA by selectively upgrading base stations with array antennas, appropriate signal processing and respective control protocols.

2.4 Cellular principle

Owing to the very limited frequency bands, a mobile radio network only has a relatively small number of speech channels available. For example, the GSM system has an allocation of 25 MHz bandwidth in the 900 MHz frequency range, which amounts to a maximum of 125 frequency channels each with a carrier bandwidth of 200 kHz. Within an eightfold time multiplex for each carrier, a maximum of 1000 channels can be realized. This number is further reduced by guardbands in the frequency spectrum and the overhead required for signaling. In order to be able to serve several hundreds of thousands or millions of subscribers in spite of this limitation, frequencies must be spatially reused, i.e. deployed repeatedly in a

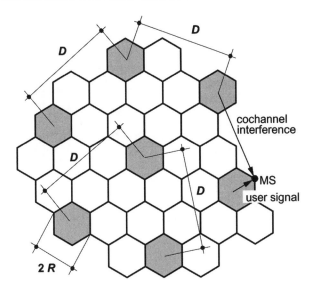

Figure 2.13 Model of a cellular network with frequency reuse.

geographic area. In this way, services can be offered with a cost-effective subscriber density and acceptable blocking probability.

2.4.1 Definitions

This spatial frequency reuse concept led to the development of the cellular principle, which allowed a significant improvement in the economic use of frequencies. The essential characteristics of the cellular network principle are as follows.

- The area to be covered is subdivided into *cells* (radio zones). These cells are often modeled in a simplified way as hexagons (Figure 2.12). The base station is located at the center of each cell.

- To each cell i a subset of the frequencies S_i is assigned from the total set (bundle) which is assigned to the respective mobile radio network. In the GSM system, the set S_i of frequencies assigned to a cell is called the Cell Allocation (CA). Two neighboring cells must never use the same frequencies, since this would lead to severe co-channel interference from the adjacent cells.

- Only at distance D (the frequency reuse distance) can a frequency from the set S_i be reused (Figure 2.13), i.e. cells with distance D to cell i can be assigned one or all of the frequencies from the set S_i belonging to cell i. When designing a mobile radio network, D must be chosen sufficiently large, such that the co-channel interference remains small enough not to affect speech quality.

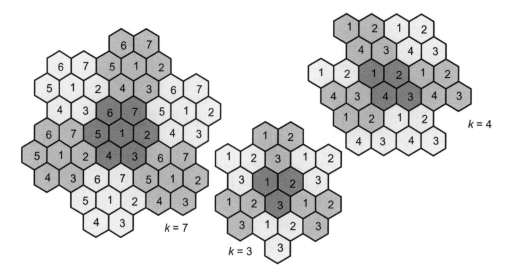

Figure 2.14 Frequency reuse and cluster formation.

- When a mobile station moves from one cell to another during an ongoing conversation, an automatic channel/frequency change occurs (*handover*), which maintains an active speech connection over cell boundaries.

The spatial repetition of frequencies is done in a regular systematic way, i.e. each cell with the cell allocation S_i sees its neighbors with the same frequencies again at a distance D (Figure 2.13). Therefore, there exist exactly six such next neighbor cells. Independent of form and size of the cells – not only in the hexagon model – the first ring in the frequency set contains six co-channel cells (see also Figure 2.14).

2.4.2   Carrier-to-interference ratio

The signal quality of a connection is measured as a function of received useful signal power and interference power received from co-channel cells and is given by the Carrier-to-Interference Ratio (CIR or C/I):

$$\frac{C}{I} = \frac{\text{Useful signal power}}{\text{Disturbing signal power}} = \frac{\text{Useful signal power}}{\text{Interference power from other cells}}. \qquad (2.14)$$

The intensity of the interference is essentially a function of co-channel interference depending on the frequency reuse distance D. From the viewpoint of a mobile station, the co-channel interference is caused by base stations at a distance D from the current base station. A worst-case estimate for the CIR of a mobile station at the border of the covered area at distance R from the base station can be obtained, subject to propagation losses, by assuming that all six neighboring interfering transmitters operate at the same power and are approximately equally far apart (a distance D that is large compared with the cell radius R)

(Lee, 1989):

$$\frac{C}{I} = \frac{P_t R^{-\gamma}}{\sum_{i=1}^{6} P_i} \approx \frac{P_t R^{-\gamma}}{\sum_{i=1}^{6} P_t D^{-\gamma}} = \frac{P_t R^{-\gamma}}{6 P_t D^{-\gamma}}, \tag{2.15}$$

where it is implied, that the mean received power according to equation (2.4) is applied, and P_i is the power received from the ith interfering cell. Finally, we find the worst-case CIR as a function of the cell radius R, the reuse distance D and the attenuation exponent γ as

$$\frac{C}{I} = \frac{R^{-\gamma}}{6 D^{-\gamma}} = \frac{1}{6} \left(\frac{R}{D} \right)^{-\gamma}. \tag{2.16}$$

Therefore, in a given radio environment, the CIR depends essentially on the ratio R/D. From these considerations it follows that for a desired or required CIR value at a given cell radius, one must choose a minimum distance for the frequency reuse, above which the co-channel interference falls below the required threshold.

2.4.3 Formation of clusters

The regular spatial repetition of frequencies results in a clustering of cells. The cells within a cluster must each be assigned different sets of channels, while cells belonging to neighboring clusters can reuse the channels in the same spatial pattern. The size of a cluster is characterized by the number of cells per cluster k, which determines the frequency reuse distance D, when the cell radius R is given. Figure 2.14 shows some examples of clusters. The numbers designate the respective frequency sets S_i used within the single cells.

For each cluster the following holds.

- A cluster can contain all of the frequencies of the mobile radio system.

- Within a cluster, no frequency can be reused. The frequencies of a set S_i may be reused at the earliest in the neighboring cluster.

- The larger a cluster is, the larger the frequency reuse distance and the larger the CIR. However, the larger the values of k, the smaller the number of channels and the number of supportable active subscribers per cell.

The frequency reuse distance D can be derived geometrically from the hexagon model depending on k and the cell radius R:

$$D = R \sqrt{3k}. \tag{2.17}$$

The CIR is then given by

$$\frac{C}{I} = \frac{R^{-\gamma}}{6 D^{-\gamma}} = \frac{R^{-\gamma}}{6 (R \sqrt{3k})^{-\gamma}} = \frac{1}{6} (3k)^{\gamma/2}. \tag{2.18}$$

Allying this result, we can now determine the optimal cluster size given a required signal quality $(C/I)_{min}$ according to the following formula:

$$\min_{i,j \geq 0} \left\{ k = i^2 + ij + j^2 \mid (C/I)_{min} \leq \frac{1}{6} (3k)^{\gamma/2} \right\}. \tag{2.19}$$

According to measurements one can assume that, for good speech understandability, a CIR of about $(C/I)_{min} = 18$ dB is sufficient. Assuming an approximate propagation coefficient of $\gamma = 4$, this yields the minimum cluster size

$$10 \log \left(\frac{C}{I} \right)_{min} = 18 \text{ dB} \iff \left(\frac{C}{I} \right)_{min} = 63.1, \tag{2.20}$$

$$\frac{1}{6}(3k)^{\gamma/2} \geq \left(\frac{C}{I} \right)_{min} = 63.1 \implies k \geq 6.5 \implies k = 7. \tag{2.21}$$

These values are also confirmed by computer simulations, which have shown that for $(C/I)_{min} = 18$ dB a reuse distance $D = 4.6R$ is required (Lee, 1989). In practically implemented networks, one can find other cluster sizes, e.g. $k = 3$ or 12.

The cellular models mentioned so far are very idealized for illustration and analysis. In reality, cells are neither circular nor hexagonal; rather they possess very irregular forms and sizes because of variable propagation conditions. An example of a possible cellular plan for a real network is shown in Figure 2.15, where one can easily recognize the individual cells with the assigned channels and the frequency reuse. The different cell sizes, which depend on whether it is an urban, suburban or rural area, are especially obvious. Figure 2.15 gives an impression of the approximate contours of equal signal power around the individual base stations. In spite of this representation, the precise fitting of signal power contours remains an idealization. The cell boundaries are, after all, blurred and defined by local thresholds, beyond which the neighboring base station's signal is received stronger than the current signal.

2.4.4 Traffic capacity and traffic engineering

As already mentioned, the number of channels and thus the maximal traffic capacity per cell depends on the cluster size k. The following relation holds:

$$n_F = \frac{B_t}{B_c k}, \tag{2.22}$$

where n_F is the number of frequencies per cell, B_t is the total bandwidth of the system and B_c is the bandwidth of one channel (including a margin for guard bands).

The number n of channels per cell in FDMA systems equals the number of frequency channels resulting from the channel and system bandwidth:

$$n = n_F. \tag{2.23}$$

The number of channels per cell in a TDMA system is the number of frequency channels multiplied by the number of time slots per channel:

$$n = m n_F, \tag{2.24}$$

where m is the number of time slots per frame.

A cell can be modeled as a traffic-theoretical loss system with n servers (channels), assuming a call arrival process with exponentially distributed interarrival times (Poisson process), and another Poisson process as a server process. Arrival and server processes are also called Markov processes, hence such a system is known as an M/M/n loss system

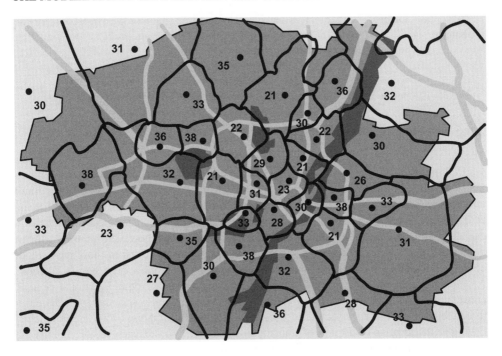

Figure 2.15 Realistic cell shapes.

(Kleinrock, 1975). For a given blocking probability B, a cell serves a maximum offered load A_{max} during the busy hour:

$$A_{max} = f(B, n) = \lambda_{max} T_m, \qquad (2.25)$$

where λ_{max} is the busy hour call attempts (BHCA) and T_m is the mean call holding time. The relation between offered load A and blocking probability B with the total number of channels n is given by the Erlang blocking formula (see Kleinrock (1975) and Tran-Gia (1996) for more details and traffic tables):

$$B = \frac{A^n/n!}{\sum_{i=0}^{n} A^i/i!}. \qquad (2.26)$$

However, these approximations are valid only for macrocellular environments, in which the number of users per cell is sufficiently large with regard to the number of available channels, such that the call arrival rate may be considered as approximately constant. For micro- and picocellular systems these assumptions usually no longer hold. Here, the traffic-theoretical dimensioning must be performed with Engset models, since the number of participants does not differ very much from the number of available channels. This results in a call arrival rate that is no longer constant. The probability that all channels are busy results from the number of users M per cell and the offer a of a non-busy user at

$$P_n = \frac{\binom{M}{n}a^n}{\sum_{i=0}^{n} \binom{M}{i}a^i}. \qquad (2.27)$$

In this case, the probability that a call arrives when no free channels are available (blocking probability) is

$$P_B = \frac{\binom{M-1}{n}a^n}{\sum_{i=0}^{n}\binom{M-1}{i}a^i}.$$ (2.28)

For $M \longrightarrow \infty$ the Engset blocking formula becomes the Erlang blocking formula.

2.4.5 Sectorization of cells

The discussion of CIR, required cluster size k and resulting traffic capacity in the previous sections indicated how an operator can design a system with a given traffic capacity per area: Since a required CIR yields a minimum cluster size k and this in turn yields a maximum number of channels per cell, the traffic capacity per area is determined by the cell radius R. Therefore, given a fixed amount of resources, i.e. a given frequency spectrum, the traffic capacity can be enhanced by choosing smaller cell sizes. This, however, requires a higher number of base stations. Therefore, the operator must not only invest in more base station equipment including backbone connections, but must also rent more base station locations. An alternative approach is the use of sectorized cells instead of omnidirectional cells, as considered so far. In the sectorized case, each omnidirectional cell is split into typically either three sectors of 120° or six sectors of 60°, each of which is supported by a sector antenna. Owing to the angular restriction of transmission and reception of each sector antenna, the number of interfering co-channel cells shrinks from six to only two co-channel cells in the case of three sectors and to just one co-channel cell in the case of six sectors. Therefore, we obtain an improved CIR of

$$\left(\frac{C}{I}\right)_{120°} \approx 3 \cdot \left(\frac{C}{I}\right)_{omni} = \frac{1}{2}\left(\frac{R}{D}\right)^{-\gamma} = \frac{1}{2}(3k)^{\gamma/2}$$ (2.29)

in the case of three sectors per cell and

$$\left(\frac{C}{I}\right)_{60°} \approx 6 \cdot \left(\frac{C}{I}\right)_{omni} = \left(\frac{R}{D}\right)^{-\gamma} = (3k)^{\gamma/2}$$ (2.30)

in the case of six sectors per cell. For a given required CIR, we can thus reduce the cluster size k to $k^{120°}$ and $k^{60°}$ for the case of three and six sectors, respectively, according to the following formulas:

$$\min_{i,j\geq0}\left\{k^{120°}=i^2+ij+j^2 \,\bigg|\, \left(\frac{C}{I}\right)_{min} \leq \frac{1}{2}(3k)^{\gamma/2}\right\}$$ (2.31)

and

$$\min_{i,j\geq0}\left\{k^{60°}=i^2+ij+j^2 \,\bigg|\, \left(\frac{C}{I}\right)_{min} \leq (3k)^{\gamma/2}\right\}.$$ (2.32)

Consequently we obtain a higher number of channels per cell. However, it is important to note that the channels assigned to each cell must now be divided among the sectors per cell. In particular, if the overall number of available channels is relatively low, this confinement of the channels to be used only within a sector leads to a reduced trunking efficiency, which can limit the capacity gain. In fact, depending on the parameters, sectorization might in extreme cases even *decrease* the capacity. The effect of sectorization on capacity is discussed in the following, using some examples.

Assuming a required CIR of $(C/I)_{min} = 18$ dB, a cluster size of $k^{omni} = 7$ is necessary (as discussed above), according to equation (2.19) with $\gamma = 4$. Using equation (2.31), we obtain $(C/I)_{k=7}^{120°} = 23$ dB for the same cluster size $k = 7$ and 120° sectorization. In the case of a reduced cluster size of $k = 4$, the resulting value is $(C/I)_{k=4}^{120°} = 18.5$ dB, which is still sufficient. Further reducing the cluster size to $k = 3$ yields $(C/I)_{k=3}^{120°} = 16$ dB, which is too low for our requirements. Thus, the cluster size can be reduced to $k^{120°} = 4$, if 120° sectorization is applied. Introducing 60° sectorization even allows for a cluster size of $k^{60°} = 3$, yielding $(C/I)_{k=3}^{60°} = 19$ dB according to equation (2.32). In order to discuss the capacity improvement, achievable with sectorization, we derive the blocking probabilities. The blocking probabilities are obtained from the Erlang-B formula, which we denote as $B(a, n)$, where a is the offered traffic in Erlang and n the number of servers, i.e. channels. In case of omnidirectional cells, the blocking probability becomes

$$P_b^{omni} = B(A, \ N^{sys}/k^{omni}), \tag{2.33}$$

where A is the offered traffic to a cell in Erlang and N^{sys} is the number of channels available to the system. In the case of sectorization, it must be accounted for the reduced trunking efficiency. Therefore, the number of channels of a single sector as well as the traffic offered to each sector must be considered. Thus, the blocking probability in the case of 120° sectorization can be expressed as

$$P_b^{120°} = B(A/3, \ N^{sys}/k^{120°}/3) \tag{2.34}$$

and for 60° sectorization we obtain

$$P_b^{60°} = B(A/6, \ N^{sys}/k^{60°}/6). \tag{2.35}$$

Using the appropriate values of k in either case, we can print the respective blocking probabilities given the number of system channels N^{sys}. Results are depicted in Figure 2.16 for $N^{sys} = 84$ and in Figure 2.17 for $N^{sys} = 252$, where in Figure 2.16 it has been considered that in the case of 60° sectorization, two sectors have four channels available, while the other four sectors can use five channels.

From the results in Figure 2.16 ($N^{sys} = 84$) it can be observed that in the case of low traffic values 60° sectorization actually performs worse than 120° sectorization and even worse than the omnidirectional system, owing to the loss in trunking efficiency with only four or five channels available to each sector. With increasing traffic load, however, 60° sectorization starts to perform better than the other two options. Assuming a desired blocking performance of $P_{b,min} = 1\%$, the capacity gain of sectorization can be obtained, which can be seen in Table 2.1. Interestingly, the gain of 60° sectorization in this case is lower than that of 120° sectorization. If the number of channels is increased ($N^{sys} = 252$), the gains in both cases increase significantly and the highest capacity is now obtained with 60° sectorization.

As a conclusion, it can be said that sectorization can increase the system capacity, however, the number of sectors should be chosen carefully depending on the number of channels available. The higher the number of channels in the system, the more we can benefit from sectorization in terms of capacity.

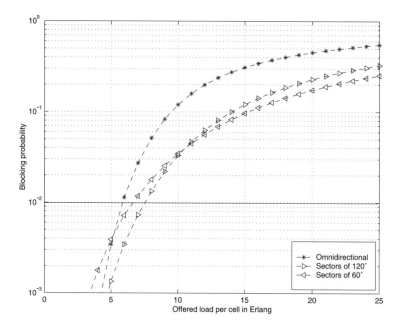

Figure 2.16   Blocking probabilities for omnidirectional and sectorized systems with 84 channels.

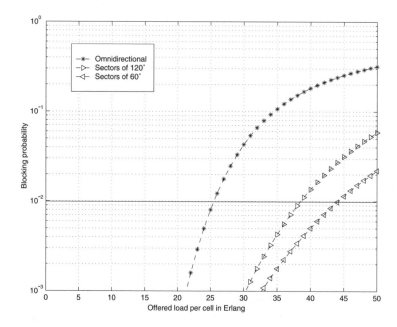

Figure 2.17   Blocking probabilities for omnidirectional and sectorized systems with 252 channels.

Table 2.1 Sectorization capacity gain with $(C/I)_{\min} = 18$ dB.

Sectors	$N^{sys} = 84$	$N^{sys} = 252$
120°	27.6%	50.6%
60°	17.8%	72.9%

2.4.6 Spatial filtering for interference reduction (SFIR)

Smart antennas at the base station can be used to perform adaptive beamforming for each individual user, yielding directional transmission on the downlink as well as directional reception on the uplink. Through this concept, the radiated and received interference power can be significantly reduced in a cellular system. One approach to apply adaptive beamforming for capacity enhancement is SDMA, as described in section 2.3.4. Another way to transform the interference reduction into actual traffic capacity gains is SFIR. While SDMA allows for intra-cellular channel reuse and has a very high potential for capacity improvement, it also demands for new advanced Radio Resource Management (RRM) methods, specifically for channel allocation and handover in order to maintain the orthogonality of spatial channels (Hartmann and Eberspächer, 2001). In contrast, SFIR does not allow for channel reuse within the cell and can thus be introduced without any modification of protocols or RRM algorithms. In an SFIR system, adaptive beams are steered towards each user; however, each user within one cell will be assigned a different traffic channel, i.e. a different FDMA or TDMA channel. This means that the user signals are separated by the orthogonal traffic channels, such that a spatial separation is not required. For this reason, intra-cell beam collisions do not have to be resolved by means of handovers. Additional RRM methods are not needed, since each traffic channel is just used once within the cell. However, how can we achieve a capacity gain with SFIR when channels are not reused within each cell? The answer is a tighter inter-cell channel reuse: owing to the reduced co-channel interference received from neighboring cells in an SFIR system, a lower reuse distance can be applied, tightening the frequency reuse and resulting in a smaller required cluster size. The lower cluster size will present the operator with a higher number of traffic channels per cell, which gives us the traffic capacity gain that we were looking for. To this end SFIR is similar to sectorization. However, there are two major differences: the advantage of SFIR as compared with sectorization is that there is no loss in terms of trunking efficiency, since channels assigned to a cell can be used, independent of the position of a user, everywhere within the cell. However, it must be considered that the interference reduction of SFIR is a statistical property. Thus, the average interference will be reduced, however, actual interference values at random time instants *may* be as high as without SFIR. The latter is the case if by chance all co-channel users are located within their cells such that their main beam pattern is steered towards the user in the center cell. In fact, the interference will fluctuate around an average value and cause outage when it surpasses a threshold value above the average value. Therefore, in order to assess the capacity of an SFIR system, we have to consider two aspects:

- average C/I improvement; and

- outage probability.

C/I improvement and blocking probability

Let us discuss the average C/I improvement first. We assume a hexagonal cell layout with six co-channel cells in the first co-channel tier. The SFIR interference scenario is depicted in Figure 2.18 for the downlink reception of a user in the center cell[1]. The antenna is modeled by a *brickwall* antenna pattern, which is defined by two parameters (Hartmann and Eberspächer, 2001): the angular width of the main lobe, denoted as β, and a constant sidelobe attenuation, denoted as D. Within the main lobe the antenna gain G is assumed to be constant as $G_0 = 1$. The probability that a user in the center cell is within the reception area of the main beam that is steered towards a user from one of the interfering cells is $p_i = \beta/2\pi$. Thus, on average, a user in the center cell will be in the main beam of a co-channel user of one of the co-channel cells $p_i \cdot 100\%$ of the time, only. Consequently, with probability $1 - p_i$ the user will not be in the main beam of a co-channel user of one of the co-channel cells. In the latter case the received interference will be attenuated by the sidelobe attenuation D produced by the beam pattern. Thus, the received mean interference in the center cell will be reduced by the factor $(p_i + (1 - p_i)/D)$. Therefore, the respective C/I value in the case of SFIR becomes

$$\left(\frac{C}{I}\right)^{\text{SFIR}} = \left(\frac{C}{I}\right)^{\text{omni}} \cdot \frac{1}{p_i + (1 - p_i)/D}, \tag{2.36}$$

where $(C/I)^{\text{omni}} = \frac{1}{6}(3k)^{\gamma/2}$ is the worst-case C/I of an omnidirectional cellular system with cluster size k and attenuation exponent γ (Eberspächer *et al.*, 2001). It is the worst-case C/I since it is based on the assumption that a channel is used in all co-channel cells at once. Likewise, we call $(C/I)^{\text{SFIR}}$, as defined in equation (2.36), the average worst-case C/I of an SFIR system. It is worst case in the sense that it is based on the assumption that a channel is used in all co-channel cells of the first co-channel tier in parallel. However, equation (2.36) represents the C/I averaged with respect to all possible beam directions of the co-channel users. With a given target $(C/I)_{\text{min}}$ we can now determine the required cluster size of our SFIR system according to

$$\min_{i,j \geq 0} \left\{ k = i^2 + ij + j^2 \,\middle|\, \left(\frac{C}{I}\right)_{\text{min}} \leq \frac{\frac{1}{6}(3k)^{\gamma/2}}{p_i + (1 - p_i)/D} \right\}. \tag{2.37}$$

Depending on the beamforming properties β and D, the resulting cluster size k^{SFIR} will be smaller than in the omnidirectional case (k^{omni}). Re-assigning the channels accordingly will result in a higher capacity. Given N^{sys} channels available to the system we have $N^{\text{omni}} = N^{\text{sys}}/k^{\text{omni}}$ channels per cell in the omnidirectional case and $N^{\text{SFIR}} = N^{\text{sys}}/k^{\text{SFIR}} = N^{\text{omni}}(k^{\text{omni}}/k^{\text{SFIR}})$ channels per cell in the SFIR case. Using the Erlang-B formula (denoted by $B(a, n)$ where a is the traffic load in Erlang offered to the cell and n the number of channels per cell), we can compute the blocking probability in either case: $P_b^{\text{omni}} = B(A, N^{\text{omni}})$ and $P_b^{\text{SFIR}} = B(A, N^{\text{SFIR}})$. However, in contrast to sectorization, the blocking probability is not sufficient to assess the SFIR capacity. In addition, the outage probability, i.e. the probability that the actual (C/I) value will be worse than the target $(C/I)_{\text{min}}$, must be considered.

---

[1] We consider the downlink case here. The uplink case is less problematic, because the base station will receive at most one interfering user as long as $\beta < 60°$ and at most two interferers if $60° \leq \beta < 120°$.

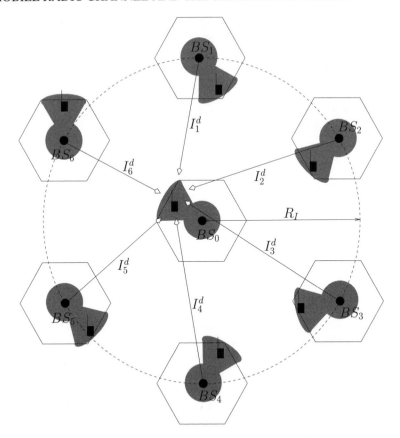

Figure 2.18 SFIR interference scenario downlink.

Outage probability

In order to determine the outage probability, we first determine a number $1 \leq n_{\mathrm{out}} \leq 6$ such that the interference threshold of the center cell user will be surpassed, if at least n_{out} of the six co-channel users have the center cell user within their main beam – after the cluster size has been reduced. In other words, n_{out} is the lowest integer value n_i, for which the C/I value in the center cell becomes lower than the threshold value $(C/I)_{\min}$ in

$$\left(\frac{C}{I}\right)^{\mathrm{omni,new}} \cdot \frac{1}{n_i/6 + (1 - n_i/6)/D} \leq \left(\frac{C}{I}\right)_{\min}, \tag{2.38}$$

with the worst case omnidirectional C/I *after* cluster size reduction: $(C/I)^{\mathrm{omni,new}} = \frac{1}{6}(3k^{\mathrm{SFIR}})^{\gamma/2}$. Thus, with the ceiling function $\lceil \cdot \rceil$, n_{out} becomes

$$n_{\mathrm{out}} = \left\lceil 6 \cdot \frac{(C/I)^{\mathrm{omni,\,new}}/(C/I)_{\min} - 1/D}{1 - 1/D} \right\rceil. \tag{2.39}$$

With n_{out} from equation (2.39) and $p_i = \beta/2\pi$ we can now determine the outage probability in case a channel is used in all six co-channel cells:

$$p(\text{out} \mid n_{\text{a}} = 6) = \sum_{n=n_{\text{out}}}^{6} \binom{6}{n} p_i^n (1 - p_i)^{6-n}, \tag{2.40}$$

where n_{a} denotes the number of active co-channel cells. Likewise, we can determine the outage probability in the more general case that a channel is used in $n_{\text{a}} \le 6$ cells of the co-channel cells:

$$p(\text{out} \mid n_{\text{a}}) = \begin{cases} \displaystyle\sum_{n=n_{\text{out}}}^{n_{\text{a}}} \binom{n_{\text{a}}}{n} p_i^n (1 - p_i)^{n_{\text{a}}-n} & n_{\text{out}} \le n_{\text{a}}, \\ 0 & n_{\text{out}} > n_{\text{a}}. \end{cases} \tag{2.41}$$

With the probabilities $p(n_{\text{a}})$ that n_{a} co-channel cells are using the same channel as the mobile in question in the target cell, we obtain the general outage probability

$$p(\text{out}) = \sum_{n_{\text{a}}=0}^{6} p(\text{out} \mid n_{\text{a}}) \cdot p(n_{\text{a}}). \tag{2.42}$$

The values of $p(n_{\text{a}})$ depend on the traffic load in the co-channel cells as well as on the number of channels available in each cell. However, they also depend on the channel assignment strategy used. Even if we assume Fixed Channel Allocation (FCA), the question remains whether channels are assigned in the same order in each cell or not. Let us assume that channels are assigned randomly, with no specific order. Now, what is the probability that a specific channel $i \in \{1, \ldots, N\}$ is active in a cell? Therefore, we define the random variable i_{a} which is 1 if channel i is active in the cell and 0 else. Since channels are assigned randomly within a cell, the probability of a channel being used in a cell is independent of the channel i:

$$p(i_{\text{a}} = 1) = \sum_{j=1}^{N} p(j \text{ channels active}) \cdot \frac{j}{N}. \tag{2.43}$$

Therefore, the probability $p(j \text{ channels active})$ that $j \le N$ channels are active in a cell is given by (Kleinrock, 1975)

$$p(j \text{ channels active}) = \frac{A^j/j!}{\sum_{k=0}^{N}(A^k/k!)}, \tag{2.44}$$

where A is the traffic load of the given cell. Thus, we obtain

$$p(i_{\text{a}} = 1) = \sum_{j=1}^{N} \frac{A^j/j!}{\sum_{k=0}^{N}(A^k/k!)} \cdot \frac{j}{N}. \tag{2.45}$$

We can now compute the probability $p(n_{\text{a}})$ that a specific channel is in use in n_{a} co-channel cells as

$$p(n_{\text{a}}) = \binom{6}{n_{\text{a}}} p(i_{\text{a}} = 1)^{n_{\text{a}}} (1 - p(i_{\text{a}} = 1))^{6-n_{\text{a}}}, \tag{2.46}$$

where we assume that the traffic load is statistically equal in each co-channel cell.

Table 2.2 SFIR capacity gain with $(C/I)_{min} = 14$ dB and $k^{SFIR} = 3$.

β	$(C/I)^{SFIR}_{k=3}$ (dB)	Capacity gain (%)	Limitation
90°	15.88	147.4	Outage
60°	17.02	268.4	Outage
45°	17.73	284.2	Blocking
30°	18.57	284.2	Blocking
20°	19.24	284.2	Blocking
15°	19.61	284.2	Blocking

With equations (2.42), (2.41), (2.46) and (2.45), we can now compute the outage probability $P_{out} = p(\text{out})$ of a user in the target cell which is using a specific channel. Given our assumption that channels are assigned in random order within each cell, this same measure is the outage probability for any active mobile in the target cell if the traffic load in each co-channel cell is A.

Numerical results

For a better understanding of the capacity of sectorized cellular networks, we discuss some numerical results in the following. We assume a target C/I value of $(C/I)_{min} = 18$ dB, yielding a cluster size of $k^{omni} = 7$. Choosing the antenna parameters $D = 10$ dB and $\beta = 30°$, we obtain $(C/I)^{SFIR}_{k=7} = (C/I)^{omni}_{k=7} + 7.6$ dB $= 25.6$ dB, when SFIR is introduced. To translate the improved C/I into capacity gain, the cluster size must be reduced. For $k = 3$ we obtain C/I values of $(C/I)^{omni}_{k=3} = 11$ dB and $(C/I)^{SFIR}_{k=3} = (C/I)^{omni}_{k=3} + 7.6$ dB $= 18.6$ dB, respectively. Since the resulting value of $(C/I)^{SFIR}_{k=3}$ is even slightly higher than the original value of $(C/I)^{omni}_{k=7}$, we could conclude that through SFIR and cluster size reduction with $k^{SFIR} = 3$ we can benefit from the capacity gain that can be achieved by channel re-arrangement. This channel re-arrangement yields $N^{SFIR} = 7/3 \cdot N^{omni}$ channels per cell if N^{omni} is the number of channels in the omnidirectional case, and for the blocking probabilities we have

$$P_b^{omni} = B(A, \, N^{omni}) \qquad (2.47)$$

and

$$P_b^{SFIR} = B(A, \, N^{omni} \cdot k^{omni}/k^{SFIR}) = B(A, \, N^{SFIR}). \qquad (2.48)$$

For $N^{omni} = 6$ channels per omnidirectional cell, yielding $N^{SFIR} = 7/3 \cdot N^{omni} = 14$ channels per SFIR cell, the respective blocking probabilities are plotted as a function of the offered traffic load per cell in Erlang in Figures 2.19, 2.20 and 2.21. As discussed above, in order to determine the SFIR capacity, also the outage probability must be considered. In Figures 2.19, 2.20 and 2.21 the outage probability is plotted for three different values of the target C/I, $(C/I)_{min} \in \{14 \text{ dB}, 16 \text{ dB}, 18 \text{ dB}\}$ and for various values of β in each case.

Let us for the moment focus on $\beta = 30°$, as chosen above. For a given target blocking probability of $P_b = 1\%$, we can obtain the carried traffic load $L_{cell}^{car,omni}$ and $L_{cell}^{car,SFIR}$ per cell in the omnidirectional case and in the SFIR case, respectively. We can now determine

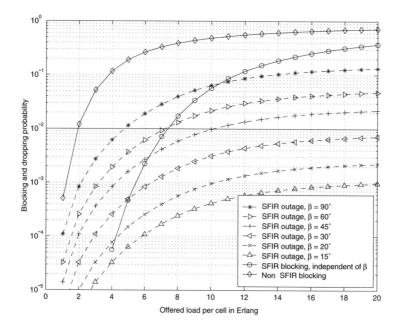

Figure 2.19 SFIR outage and blocking with $(C/I)_{min} = 14$ dB and $k^{SFIR} = 3$.

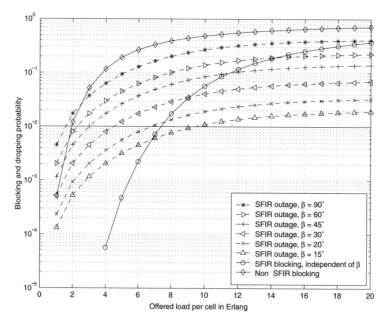

Figure 2.20 SFIR outage and blocking with $(C/I)_{min} = 16$ dB and $k^{SFIR} = 3$.

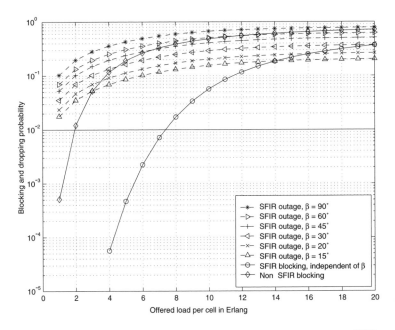

Figure 2.21 SFIR outage and blocking with $(C/I)_{\min} = 18$ dB and $k^{\text{SFIR}} = 3$.

Table 2.3 SFIR capacity gain with $(C/I)_{\min} = 16$ dB and $k^{\text{SFIR}} = 3$.

β	$(C/I)^{\text{SFIR}}_{k=3}$ (dB)	Capacity gain (%)	Limitation
90°	15.88	−21.1	Outage
60°	17.02	21.1	Outage
45°	17.73	57.9	Outage
30°	18.57	131.6	Outage
20°	19.24	257.9	Outage
15°	19.61	284.2	Blocking

Table 2.4 SFIR capacity gain with $(C/I)_{\min} = 18$ dB and $k^{\text{SFIR}} = 3$.

β	$(C/I)^{\text{SFIR}}_{k=3}$ (dB)	Capacity gain	Limitation
90°	15.88	<0	Outage
60°	17.02	<0	Outage
45°	17.73	<0	Outage
30°	18.57	<0	Outage
20°	19.24	<0	Outage
15°	19.61	<0	Outage

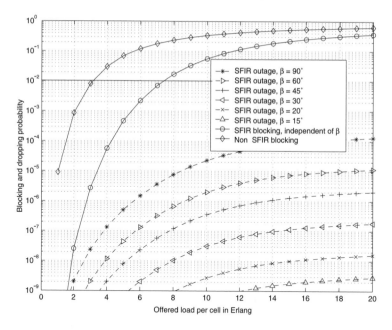

Figure 2.22 SFIR outage and blocking with $(C/I)_{\min} = 14$ dB and $k^{SFIR} = 4$.

Table 2.5 SFIR capacity gain with $(C/I)_{\min} = 14$ dB and $k^{SFIR} = 4$.

β	$(C/I)^{SFIR}_{k=4}$ (dB)	Capacity gain (%)	Limitation
90°	18.68	128.1	Blocking
60°	19.82	128.1	Blocking
45°	20.53	128.1	Blocking
30°	21.37	128.1	Blocking
20°	22.04	128.1	Blocking
15°	22.42	128.1	Blocking

the SFIR capacity gain $G^{SFIR} = (L^{car,SFIR}_{cell}/L^{car,omni}_{cell} - 1) \cdot 100\%$ and we obtain $G^{SFIR} = 284.2\%$, as long as we are only concerned with the blocking probability ($L^{car,SFIR}_{cell}$ and $L^{car,omni}_{cell}$ are the carried traffic load in the SFIR case and in the omni case, respectively). Now we have to choose a target outage probability. Note that this kind of outage, as opposed to outage caused by fast fading, will last in the order of seconds at least, depending on the user mobility. Since this can easily cause a call to be dropped, we choose the target outage probability to be $P_{out} = 1\%$. To see whether the outage probability constraint will put a more severe limit on the carried traffic load as the blocking probability, we observe from the plots at which traffic load the outage probability crosses the 1% line. From Figure 2.19 we see

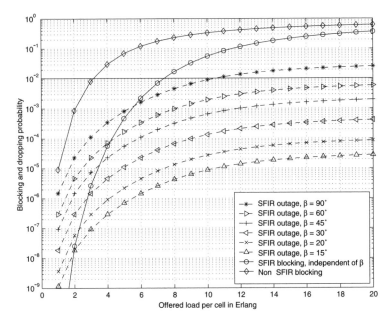

Figure 2.23 SFIR outage and blocking with $(C/I)_{min} = 16$ dB and $k^{SFIR} = 4$.

Table 2.6 SFIR capacity gain with $(C/I)_{min} = 16$ dB and $k^{SFIR} = 4$.

β	$(C/I)^{SFIR}_{k=4}$ (dB)	Capacity gain (%)	Limitation
90°	18.68	128.1	Blocking
60°	19.82	128.1	Blocking
45°	20.53	128.1	Blocking
30°	21.37	128.1	Blocking
20°	22.04	128.1	Blocking
15°	22.42	128.1	Blocking

that in the case of $(C/I)_{min} = 14$ dB the outage constraint becomes a limiting factor for a higher load than the blocking constraint. Thus, in this case the SFIR capacity is blocking limited and the above capacity gain of 284.2% can be fully utilized. However, moving on to the case of $(C/I)_{min} = 16$ dB, we see from Figure 2.20 that the outage constraint becomes more severe than the blocking constraint, which reduces the carried traffic load and thus the capacity gain shrinks to 131.6%. Finally, looking at Figure 2.21 we observe that the outage constraint leads to a lower traffic load than in the omnidirectional case. Thus, in the case of $(C/I)_{min} = 18$ dB we actually obtain a reduced capacity through SFIR, unless we accept a higher outage probability.

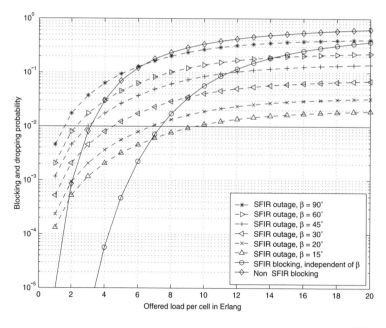

Figure 2.24 SFIR outage and blocking with $(C/I)_{min} = 18$ dB and $k^{SFIR} = 4$.

Table 2.7 SFIR capacity gain with $(C/I)_{min} = 18$ dB and $k^{SFIR} = 4$.

β	$(C/I)^{SFIR}_{k=4}$ (dB)	Capacity gain (%)	Limitation
90°	18.68	−50.0	Outage
60°	19.82	−28.1	Outage
45°	20.53	−6.3	Outage
30°	21.37	40.6	Outage
20°	22.04	112.5	Outage
15°	22.42	128.1	Blocking

Tables 2.2, 2.3 and 2.4 contain the achieved capacity gains for a variety of values of β for each of the target C/I values. We can conclude that the capacity gain decreases with growing requirements on the signal quality (growing $(C/I)_{min}$), and the capacity gain increases as the antenna main lobe becomes narrower. It is also indicated in the tables whether the capacity is rather limited by the blocking constraint or by the outage constraint. Once the system becomes blocking limited, the capacity is independent of the beamwidth.

Looking at the results in Table 2.4, we see that for $(C/I)_{min} = 18$ dB, the capacity actually decreases with SFIR for all considered values of β. This is remarkable, since according to the averaged worst-case C/I, which is $(C/I)^{SFIR}_{k=3} = 18.6$ dB for $\beta = 30°$, as computed above, a cluster size reduction from $k = 7$ down to $k = 3$ appears to be feasible.

However, the outage constraint limits the capacity so severely, that it becomes actually lower than in the omnidirectional case. Therefore, we include the results for the case that the cluster size is reduced to $k^{\mathrm{SFIR}} = 4$ instead of $k^{\mathrm{SFIR}} = 3$. Also the number of available channels is altered to $N^{\mathrm{omni}} = 8$ resulting in $N^{\mathrm{SFIR}} = 7/4 \cdot N^{\mathrm{omni}} = 14$ channels per cell after channel reassignment. The results are shown in Figures 2.22, 2.23 and 2.24 as well as in Tables 2.5, 2.6 and 2.7. The results shown in Table 2.7 show that now, with a cluster size of $k = 4$, a capacity gain of 40.6% is achieved for $\beta = 30°$ and $(C/I)_{\mathrm{min}} = 18$ dB. Since the reuse distance is higher in this case, the outage constraint does not limit the capacity as strongly.

3

System architecture and addressing

3.1 System architecture

The fundamental components of a GSM network are shown in Figure 3.1. A user carries a Mobile Station (MS), which can communicate over the air with a base station, called Base Tranceiver Station (BTS) in GSM. The BTS contains transmitter and receiver equipment, such as antennas and amplifiers, as well as a few components for signal and protocol processing. For example, error protection coding is performed in the BTS, and the link-level protocol for signaling on the radio path is terminated here. In order to keep the base stations small, the essential control and protocol intelligence resides in the Base Station Controller (BSC). It contains, for example, protocol functions for radio channel allocation, channel setup and management of handovers. Typically, several BTSs are controlled by one BSC. In practice, the BTS and BSC are connected by fixed lines or point-to-point radio links. BTS and BSC together form the radio access network.

The combined traffic of the users is routed through a switch, called the Mobile Switching Center (MSC). It performs all of the switching functions of a switching node in a fixed telephone network, e.g., in an Integrated Services Digital Network (ISDN). This includes path search, data forwarding and service feature processing. The main difference between an ISDN switch and an MSC is that the MSC also has to consider the allocation and administration of radio resources and the mobility of the users. The MSC therefore has to provide additional functions for location registration of users and for the handover of a connection in the case of changing from cell to cell. A cellular network can have several MSCs with each being responsible for a part of the network (e.g., a city or metropolitan area). Calls originating from or terminating in the fixed network are handled by a dedicated Gateway MSC (GMSC). The interworking of a cellular network and a fixed network (e.g., PSTN, ISDN) is performed by the Interworking Function (IWF). It is needed to map the protocols of the cellular network onto those of the respective fixed network. Connections to other mobile or international networks are typically routed over the International Switching Center (ISC) of the respective country.

GSM – Architecture, Protocols and Services Third Edition   J. Eberspächer, H.-J. Vögel, C. Bettstetter and C. Hartmann
© 2009 John Wiley & Sons, Ltd

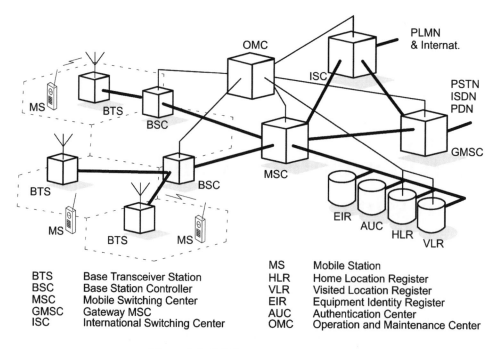

BTS	Base Transceiver Station	MS	Mobile Station
BSC	Base Station Controller	HLR	Home Location Register
MSC	Mobile Switching Center	VLR	Visited Location Register
GMSC	Gateway MSC	EIR	Equipment Identity Register
ISC	International Switching Center	AUC	Authentication Center
		OMC	Operation and Maintenance Center

Figure 3.1 GSM system architecture.

A GSM network also contains several types of databases. The Home Location Register (HLR) and the Visited Location Register (VLR) store the current location of a mobile user. This is needed since the network must know the current cell of a user to establish a call to the correct base station. In addition, these registers store the profiles of users, which are required for charging and billing and other administrative issues. Two further databases perform security functions: the Authentication Center (AUC) stores security-related data such as keys used for authentication and encryption; the Equipment Identity Register (EIR) registers equipment data rather than subscriber data.

The network management is organized from a central place, the Operation and Maintenance Center (OMC). Its functions include the administration of subscribers, terminals, charging data, network configuration, operation, performance monitoring and network maintenance. The operation and maintenance functions are based on the concept of the Telecommunication Management Network (TMN) which is standardized in the ITU-T series M.30.

In summary, a GSM network can be divided into three subnetworks: the radio access network, the core network and the management network. These subnetworks are called subsystems in the GSM standard. The respective three subsystems are called the Base Station Subsystem (BSS), the Network Switching Subsystem (NSS) and the Operation and Maintenance Subsystem (OMSS).

Figure 3.2 summarizes the hierarchical relationship between the network components MSC, BSC and BTS. The entire network is divided into MSC regions. Each of these is composed of at least one Location Area (LA), which in turn consists of several cell groups.

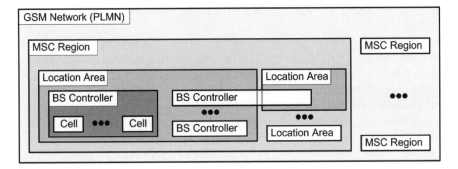

Figure 3.2 GSM system hierarchy.

Figure 3.3 Mobile equipment personalization with the SIM.

Each cell group is assigned to a BSC. For each LA there exists at least one BSC, but cells of one BSC may belong to different LAs. The exact partitioning of the network area with respect to LAs, BSCs and MSCs is not, however, uniquely determined and is left to the network operator who thus has many possibilities for optimization.

3.2   The SIM concept

Each GSM user owns a personal chip card, the Subscriber Identity Module (SIM). As illustrated in Figure 3.3, it can be plugged into a piece of mobile equipment. In fact, only the SIM of a subscriber turns a piece of mobile equipment into a complete mobile station with network usage privileges, which can be used to make calls or receive calls.

   This concept allows us to distinguish between equipment mobility and subscriber mobility. The subscriber can register to the locally available network with their SIM card on different mobile stations, or the SIM card could be used as a normal telephone card in the fixed telephone network. This enables international roaming independent of mobile equipment and network technology, provided that the interface between SIM and end terminal is standardized.

Beyond that, the SIM can store short messages and charging information, and it has a telephone book function and short list of call numbers storing names and telephone numbers for efficient and fast number selection. These functions, in particular, contribute to a genuine personalization of a mobile terminal, since the subscriber can use their normal 'environment' plus telephone list and short message archive with any piece of mobile equipment. In addition to subscriber-specific data, the SIM can also store network-specific data, e.g., lists of carrier frequencies used by the network to broadcast system information periodically. Use of the SIM and thus of the whole MS can be protected with a Personal Identification Number (PIN) against unauthorized access.

The SIM also takes over security functions: all of the cryptographic algorithms to be kept confidential are realized on the SIM, which implements important functions for the authentication and user data encryption based on the subscriber identity and secret keys.

3.3   Addressing

As in each communication network, the entities of a GSM network must be assigned certain addresses or identities. These serve to identify, authenticate and localize the network entities. The most commonly known GSM address is the telephone number of a user. In addition to telephone numbers, several other identifiers have been defined; they are needed for the management of user mobility and for addressing all remaining network elements.

GSM distinguishes explicitly between a user and their equipment. Hence, there are specific address types for users and specific address types for MSs. The user identities are stored on the SIM; the equipment identities on the mobile equipment. In addition, GSM distinguishes between user identity and their telephone number. This leaves some scope for development of services when each subscriber may be called personally, independent of reachability or type of connection (mobile or fixed). In addition to the personal identifier, each GSM subscriber is assigned one or several ISDN numbers. The following sections explain the most important addresses and identifiers used in GSM.

3.3.1   International mobile station equipment identity

The International Mobile Station Equipment Identity (IMEI) uniquely identifies a mobile station internationally and gives clues about its manufacturer and the date of manufacturing. It is a kind of serial number. The IMEI is allocated by the equipment manufacturer and registered by the network operator, who stores it in the EIR.

By means of the IMEI one recognizes obsolete, stolen or nonfunctional equipment and can deny service if required. For this purpose, the IMEI is assigned to one or more of the following three categories within the EIR.

- The white list is a register of all equipment.

- The black list contains all suspended equipment. This list is periodically exchanged among network operators.

- Optionally, an operator may maintain a gray list, in which malfunctioning equipment or equipment with obsolete software versions is registered. Such equipment has network access, but its use is reported to the operating personnel.

The IMEI is usually requested by the network at registration, but it can be requested repeatedly. It is a hierarchical address, containing the following parts:

- Type Approval Code (TAC), six digits, centrally assigned;

- Final Assembly Code (FAC), six digits, assigned by the manufacturer;

- Serial Number, six digits, assigned by the manufacturer;

- Spare, one digit.

3.3.2 International mobile subscriber identity

When registering for service with a mobile network operator, each subscriber receives a unique identifier, the International Mobile Subscriber Identity (IMSI). This IMSI is stored in the SIM. A mobile station can only be operated if a SIM with a valid IMSI is inserted into equipment with a valid IMEI, since this is the only way to correctly bill the associated subscriber. The IMSI uses a maximum of 15 decimal digits and consists of three parts:

- Mobile Country Code (MCC), three digits, internationally standardized;

- Mobile Network Code (MNC), two digits, for unique identification of mobile networks within a country;

- Mobile Subscriber Identification Number (MSIN), maximum of 10 digits, identification number of the subscriber in their mobile home network.

The IMSI is a GSM-specific addressing concept and is different from the ISDN numbering plan. A three-digit MCC has been assigned to each of the GSM countries, and two-digit MNCs have been assigned within countries (e.g., 262 as MCC for Germany; and MNC 01, 02 and 07 for the networks of T-Mobile, Vodafone, and O2, respectively). Whereas the MCC is defined internationally, the National Mobile Subscriber Identity (NMSI = MNC + MSIN) is assigned by the operator of the home network.

3.3.3 Mobile subscriber ISDN number

The 'real telephone number' of a mobile user is called the Mobile Subscriber ISDN Number (MSISDN). It is assigned to the subscriber (their SIM), such that a mobile station can have several MSISDNs depending on the SIM. With this concept, GSM was the first mobile system to distinguish between subscriber identity and the number to call. The separation of call number (MSISDN) and subscriber identity (IMSI) primarily serves to protect the confidentiality of the IMSI. In contrast to the MSISDN, the IMSI need not be made public. With this separation, one cannot derive the subscriber identity from the MSISDN, unless the association of IMSI and MSISDN as stored in the HLR has been made public. It is the rule that the IMSI used for subscriber identification is not known, and thus the faking of a false identity is significantly more difficult.

In addition, a subscriber can hold several MSISDNs for selection of different services. Each MSISDN of a subscriber is reserved for specific service (voice, data, fax, etc.). In order to realize this service, service-specific resources have to be activated in the MS as well as in the network. The service desired and the resources needed for the specific call can be derived from the MSISDN. Thus, an automatic activation of service-specific resources is already possible during the setup of a connection. The MSISDN categories follow the international ISDN numbering plan, having the following structure:

- Country Code (CC), up to three digits;

- National Destination Code (NDC), typically two or three digits;

- Subscriber Number (SN), a maximum of 10 digits.

The CCs are internationally standardized, complying with the ITU-T recommendation E.164. There are country codes with one, two, or three digits, e.g. the country code for the USA is 1, for the UK it is 44 and for Finland it is 358. The national operator or regulatory administration assigns the NDC as well as the SN, which may have variable length. The NDC of the mobile networks in Germany have three digits (e.g., 170, 171, 172). The MSISDN is stored centrally in the HLR.

3.3.4   Mobile station roaming number

The Mobile Station Roaming Number (MSRN) is a temporary location-dependent ISDN number. It is assigned by the locally responsible VLR to each MS in its area. Calls are routed to the MS by using the MSRN. On request, the MSRN is passed from the HLR to the GMSC. The MSRN has the same structure as the MSISDN:

- CC of the visited network;

- NDC of the visited network;

- SN in the current mobile network.

The components CC and NDC are determined by the network visited and depend on the current location. The SN is assigned by the current VLR and is unique within the mobile network. An MSRN is assigned in such a way that the currently responsible switching node MSC in the visited network can be determined from the subscriber number, which allows routing decisions to be made.

The MSRN can be assigned in two different ways by the VLR: either at each registration when the MS enters a new LA or each time when the HLR requests it for setting up a connection for incoming calls to the MS.

In the first case, the MSRN is also passed on from the VLR to the HLR, where it is stored for routing. In the case of an incoming call, the MSRN is first requested from the HLR of this MS. In this way the currently responsible MSC can be determined, and the call can be routed to this switching node. Additional localization information can be obtained there from the responsible VLR.

In the second case, the MSRN cannot be stored in the HLR, since it is only assigned at the time of call setup. Therefore, the address of the current VLR must be stored in the tables of the HLR. Once routing information is requested from the HLR, the HLR itself goes to the current VLR and uses a unique subscriber identification (IMSI and MSISDN) to request a valid roaming number MSRN. This allows further routing of the call.

3.3.5 Location area identity

Each LA of a cellular network has its own identifier. The Location Area Identifier (LAI) is also structured hierarchically and internationally unique, with LAI again consisting of an internationally standardized part and an operator-dependent part:

- CC, three digits;

- MNC, two digits;

- Location Area Code (LAC), a maximum of five digits or a maximum of 2×8 bits, coded in hexadecimal.

This LAI is broadcast regularly by the base station on the Broadcast Control Channel (BCCH). Thus, each cell is identified uniquely on the radio channel as belonging to an LA, and each MS can determine its current location through the LAI. If the LAI that is 'heard' by the MS changes, the MS notices this LA change and requests an update to its location information in the VLR and HLR (location update). The significance for GSM networks is that the MS itself rather than the network is responsible for monitoring the local conditions of signal reception, to select the base station that can be received best, and to register with the VLR of that LA which the current base station belongs to. The LAI is requested from the VLR if the connection for an incoming call has been routed to the current MSC using the MSRN. This determines the precise location of the MS where the mobile can be subsequently paged. When the MS answers, the exact cell and therefore also the base station become known; this information can then be used to switch the call through.

3.3.6 Temporary mobile subscriber identity

The VLR responsible for the current location of a subscriber can assign a Temporary Mobile Subscriber Identity (TMSI), which has only local significance in the area handled by the VLR. It is used in place of the IMSI for the definite identification and addressing of the MS. In this way nobody can determine the identity of the subscriber by listening to the radio channel, since this TMSI is only assigned during the presence of the MS in the area of one VLR, and can even be changed during this period (ID hopping). The MS stores the TMSI on the SIM card. The TMSI is stored on the network side only in the VLR and is not passed to the HLR. A TMSI may therefore be assigned in an operator-specific way; it can consist of up to 4×8 bits, but the HEX value FFFF FFFF is excluded, because the SIM marks empty fields internally with logical 1.

Together with the current location area, a TMSI allows a subscriber to be identified uniquely, i.e., for the ongoing communication the IMSI is replaced by the 2-tuple (TMSI, LAI).

3.3.7   Other identifiers

The VLR can assign an additional searching key to each MS within its area to accelerate database access; this is the Local Mobile Station Identity (LMSI). The LMSI is assigned when the MS registers with the VLR and is also sent to the HLR. The LMSI is no longer used by the HLR, but each time messages are sent to the VLR concerning a MS, the LMSI is added, so the VLR can use the short searching key for transactions concerning this MS. This kind of additional identification is only used when the MSRN is newly assigned with each call. In this case, fast processing is very important to achieve short times for call setup. Like the TMSI, an LMSI is also assigned in an operator-specific way, and it is only unique within the administrative area of a VLR. An LMSI consists of four octets (4×8 bits).

Within an LA, the individual cells are uniquely identified with a Cell Identifier (CI), a maximum of 2×8 bits. Together with the global CI cells are thus also internationally defined in a unique way.

In order to distinguish neighboring base stations, these receive a Base Transceiver Station Identity Code (BSIC) which consists of two components:

- Network Color Code (NCC), a color code within a mobile network (3 bits);

- Base Transceiver Station Color Code (BCC), a BTS color code (3 bits).

The BSIC is broadcast periodically by the base station. Directly adjacent mobile networks must have different NCCs, and neighboring base stations of a mobile network must have different BCCs.

MSCs and location registers (HLR, VLR) are addressed with ISDN numbers. In addition, they may have a Signaling Point Code (SPC) within a mobile network, which can be used to address them uniquely within the Signaling System Number 7 network (SS#7). The number of the VLR in whose area a MS is currently roaming must be stored in the HLR data for this MS, if the MSRN distribution is on a call-by-call basis; thus the MSRN can be requested for incoming calls and the call can be switched through to the MS.

3.4   Registers and subscriber data

3.4.1   Location registers (HLR and VLR)

The GSM standard defines two database types for the management of user data and location: the HLR and the VLR. These databases are queried by the network for user registration and localization.

The HLR has a record for all subscribers registered with a network operator. It stores, for example, each user's telephone number, service subscriptions, permissions and authentication data. In addition to this permanent administrative data, it also contains temporary data, such as the current location of a subscriber. In the case of incoming traffic to a mobile user, the HLR is queried to determine the user's current location. This enables the gateway to route the traffic to the appropriate MSC. The MS must inform the network about its current location area; to do so it sends a location update message to the network whenever it changes its location area. The full list of subscriber data stored in the HLR is given in Table 3.1.

A VLR is responsible for a group of location areas and stores the data of all users that are currently located in this area. The data includes part of the permanent user data, which is

Table 3.1 Mobile subscriber data in the HLR.

Subscriber and subscription data:	– IMSI – MSISDN – Service subscriptions – Service restrictions (e.g., roaming restrictions) – Information on the subscriber's equipment (if available) – Authentication data (subject to implementation)
Tracking and routing information:	– Mobile Station Roaming Number (MSRN) – Current VLR address (if available) – Current MSC address (if available) – Local Mobile Subscriber Identity (LMSI) (if available)

Table 3.2 Mobile subscriber data in the VLR.

Subscriber and subscription data:	– IMSI – MSISDN – Parameters for supplementary services – Information on the subscriber's equipment (if available) – Authentication data (subject to implementation)
Tracking and routing information:	– MSRN – TMSI – LMSI (if available) – LAI of LA where the MS was registered (used for paging and call setup)

copied from the HLR to the VLR for fast access. In addition, the VLR may also assign and store local data, such as temporary identifiers. A user may either be registered with a VLR of their home network or a foreign network. Upon a location update, the MSC forwards the user's identity and current location to the VLR, which subsequently updates its database. If the user has not been registered with this VLR before, the HLR is informed about the current VLR of the user. This process enables incoming calls to be routed to this MS. Table 3.2 summarizes the subscriber data stored in the VLR.

Typically, there is one central HLR per network and one VLR for each MSC. This organization depends on the number of subscribers, the processing and storage capacity of the switches and the structure of the network.

3.4.2   Security-related registers (AUC and EIR)

Two additional databases are responsible for various aspects of system security. System security of GSM networks is based primarily on the verification of equipment and subscriber

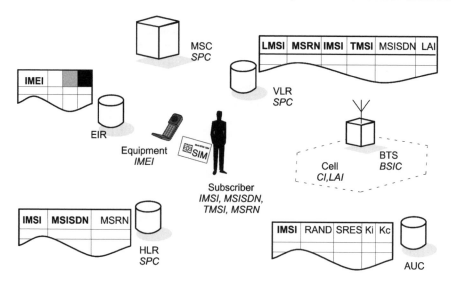

Figure 3.4 GSM databases and addresses.

identity; therefore, the databases serve for subscriber identification and authentication and for equipment registration. Confidential data and keys are stored or generated in the AUC. The keys serve for user authentication and authorize the respective service access. The EIR stores the serial numbers (supplied by the manufacturer) of the terminals (IMEI), which makes it possible to check for MSs with obsolete software or to block service access for MSs reported as stolen.

3.4.3   Subscriber data

Service-specific data are used to parameterize and personalize supplementary services. Finally, contracts with subscribers can define different service levels, e.g., booking of special supplementary services or subscriptions to data or teleservices. The contents of such contracts are stored in appropriate data structures in order to enable correct realization or provision of these services.

The association of the most important identifiers and their storage locations is summarized in Figure 3.4. Subscriber-related addresses are stored on the SIM and in the HLR and VLR as well. These data (IMSI, MSISDN, TMSI, MSRN) serve to address, identify and localize a subscriber or a MS. Whereas IMSI and MSISDN are permanent data items, TMSI and MSRN are temporary values, which change according to the current location of the subscriber. Of the other data items defined for user or network equipment elements (such as IMEI, LAI or SPCs), only some are used (LAI, SPC) for localizing or routing. IMEI and BSIC/CI hold a special position by being used only for identification of network elements.

Security-relevant subscriber data are stored in the AUC, which also calculates identifiers and keys for cryptographic processing functions. Each set of data in the AUC contains the IMSI of the subscriber as a search key. For identification and authentication of a subscriber,

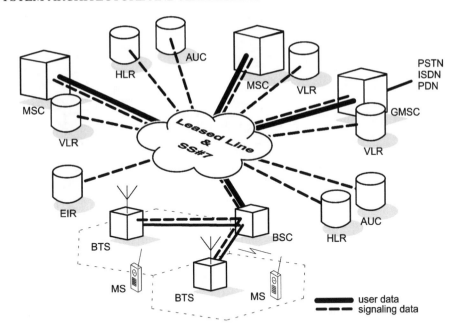

Figure 3.5 User data transport and signaling in a GSM network.

the AUC stores the subscriber's secret key Ki from which a pair of keys RAND/SRES are precalculated and stored. Once an authentication request occurs, this pair of keys is queried by the VLR to conduct the identification/authentication process properly. The key Kc for user data encryption on the radio channel is also calculated in advance in the AUC from the secret key Ki and is requested by the VLR at connection setup.

Above all, the HLR contains the permanent data about the subscriber's contractual relationship, e.g., information about subscribed bearer and teleservices (data, fax, etc.), service restrictions and parameters for supplementary services. Beyond that, the registers also contain information about equipment used by the subscriber (IMEI). Depending on the implementation of the authentication center and the security mechanisms, data and keys used for subscriber authentication and encryption can also be stored there. The search keys used for retrieving subscriber information (such as IMSI, MSISDN, MSRN, TMSI and LMSI) from a register are indicated in boldface (Figure 3.4).

3.5   Network interfaces and configurations

Figure 3.5 shows the GSM system architecture with corresponding user data and signaling links between the network components.

Signaling has two fundamentally different parts: The core network employs the SS#7, which is well known from fixed networks. In order to setup, manage and release calls, the SS#7 protocol called ISDN User Part (ISUP) is used. In order to perform signaling that

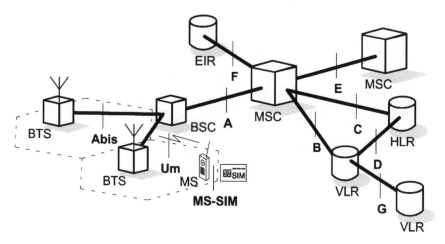

Figure 3.6 Interfaces in a GSM network.

is specific to mobile networking, an extension to SS#7 has been developed, the so-called Mobile Application Part (MAP). It is implemented in the MSC, HLR and VLR.

The radio access network (including the air interface) does not employ the SS#7 protocol, but uses a GSM-specific protocol. Signaling between the radio access network and the MSC uses the Base Station System Application Part (BSSAP).

3.5.1   Interfaces

The communication relationships between the GSM network components are formally described by a number of standardized interfaces (Figure 3.6).

The A interface between BSS and MSC is used for the transfer of data for BSS management, for connection control and for mobility management. Within the BSS, the Abis interface between BTS and BSC and the air interface Um have been defined.

An MSC which needs to obtain data about an MS staying in its administrative area, requests the data from the VLR responsible for this area over the B interface. Conversely, the MSC forwards to this VLR any data generated at location updates by MSs. If the subscriber reconfigures special service features or activates supplementary services, the VLR is also informed first, which then updates the HLR.

This updating of the HLR occurs through the D interface. The D interface is used for the exchange of location-dependent subscriber data and for subscriber management. The VLR informs the HLR about the current location of the mobile subscriber and reports the current MSRN. The HLR transfers all of the subscriber data to the VLR that is needed to give the subscriber their usual customized service access. The HLR is also responsible for giving a cancellation request for the subscriber data to the old VLR once the acknowledgement for the location update arrives from the new VLR. If, during location updating, the new VLR needs data from the old VLR, it is directly requested over the G interface. Furthermore, the identity of subscriber or equipment can be verified during a location update; for requesting and checking the equipment identity, the MSC has an interface F to the EIR.

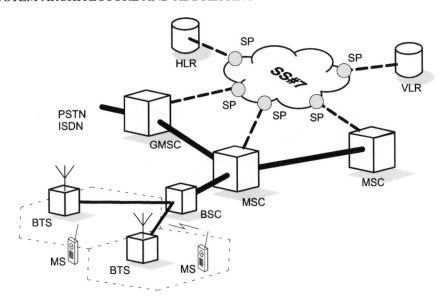

Figure 3.7 Basic configuration of a GSM network.

An MSC has two more interfaces in addition to the A and B interfaces, namely the C and E interfaces. Charging information can be sent over the C interface to the HLR. In addition to this, the MSC must be able to request routing information from the HLR during call setup, for calls from the mobile network as well as for calls from the fixed network. In the case of a call from the fixed network, if the fixed network's switch cannot interrogate the HLR directly, initially it routes the call to a GMSC, which then interrogates the HLR. If the mobile subscriber changes during a conversation from one MSC area to another, a handover needs to be performed between these two MSCs, which occurs across the E interface.

3.5.2 Configurations

As mentioned above, the configuration of a mobile network is largely left to the network operator. Figure 3.7 shows a basic configuration of a GSM network. This configuration contains a central HLR and a central VLR. All database transactions (updates, inquiries, etc.) and handover transactions between the MSC are performed with the help of the MAP over the SS#7 network. For this purpose, each MSC and register is known as a Signaling Point (SP) and is identified by its Signaling Point Code (SPC) within the SS#7 network. Whenever an MS changes its location area, the location information in the VLR must be updated. Furthermore, the VLR has to be interrogated: the MSC needs subscriber parameters in addition to location data for successful connection setup, such as service restrictions and supplementary services to be activated. Thus, there is a significant message traffic between MSC and VLR, which constitutes an ensuing load on the signaling network.

Hence, these two functional units can be combined into one physical unit, i.e. the entire VLR is implemented in distributed form and a VLR is associated with each MSC

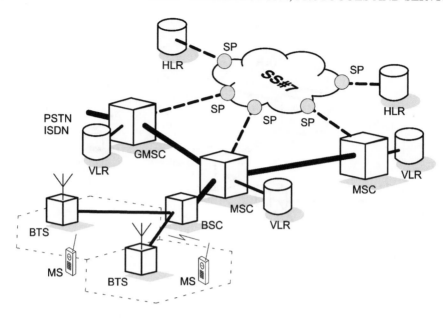

Figure 3.8 Configuration of a GSM network with a VLR for each MSC.

(see Figure 3.8). The traffic between MSC and VLR no longer needs to be transported through the SS#7 network.

We can go one step further and also distribute the database of the HLR, thus introducing several HLRs into a mobile network. This is especially interesting for a growing pool of subscribers, since a centralized database leads to a high traffic load for this database. If there are several HLRs in a network, the network operator has to define an association rule between MSISDNs and HLRs, such that for incoming calls the routing information to an MSISDN can be derived from the associated HLR. One possible association is geographic partitioning of the whole subscriber identification space (SN field in the MSISDN), where, for example, the first two digits of the SN indicate the region and the associated HLR. In extreme cases, the HLR can be realized with the VLR in a single physical unit. In this case, an HLR would also be associated with each MSC.

4

Air interface – physical layer

The GSM physical layer, which resides on the first of the seven layers of the Open Systems Interconnection (OSI) Reference Model (Tanenbaum, 1996), contains very complex functions. The physical channels are defined here by a TDMA scheme. On top of the physical channels, a series of logical channels are defined, which are transmitted in the time slots of the physical channels. Logical channels perform a multiplicity of functions, such as payload transport, signaling, broadcast of general system information, synchronization and channel assignment.

The structure of this chapter is as follows: In section 4.1, we describe the logical channels. This serves as a foundation for understanding the signaling procedures at the air interface. The realization of the physical channels, including GSM modulation, multiple access, duplexing and frequency hopping follows in section 4.2. Next, section 4.3 covers synchronization. The mapping of logical onto physical channels follows in section 4.4, where the higher-level multiplexing of logical channels into multiframes is also covered. Section 4.5 contains a discussion of the most important control mechanisms for the air interface (channel measurement, power control, disconnection and cell selection). The conclusion of the chapter is a power-up scenario with the sequence of events occurring, from when a MS is turned on to when it is in a synchronized state ready to transmit (section 4.9).

4.1 Logical channels

On Layer 1 of the OSI Reference Model, GSM defines a series of logical channels, which are made available either in an unassigned random access mode or in a dedicated mode assigned to a specific user. Logical channels are divided into two categories (Table 4.1): traffic channels and signaling (control) channels.

4.1.1 Traffic channels

The Traffic Channels (TCHs) are used for the transmission of user payload data (speech, data). They do not carry any control information of Layer 3. Communication over a TCH can be circuit-switched or packet-switched. In the circuit-switched case, the TCH provides a transparent data connection or a connection that is specially treated according to the carried

Table 4.1 Classification of logical channels in GSM.

Group		Channel	Function	Direction
Traffic channel	TCH	TCH/F, Bm	Full-rate TCH	MS ↔ BSS
		TCH/H, Lm	Half-rate TCH	MS ↔ BSS
Signaling channels (Dm)	BCH	BCCH	Broadcast control	MS ← BSS
		FCCH	Frequency correction	MS ← BSS
		SCH	Synchronization	MS ← BSS
	CCCH	RACH	Random access	MS → BSS
		AGCH	Access grant	MS ← BSS
		PCH	Paging	MS ← BSS
		NCH	Notification	MS ← BSS
	DCCH	SDCCH	Stand-alone dedicated control	MS ↔ BSS
		SACCH	Slow associated control	MS ↔ BSS
		FACCH	Fast associated control	MS ↔ BSS

service (e.g. telephony). For the packet-switched mode, the TCH carries user data of OSI Layers 2 and 3 according to the recommendations of the X.25 standard or similar standard packet protocols.

A TCH may either be fully used (full-rate TCH, TCH/F) or be split into two half-rate channels (half-rate TCH, TCH/H), which can be allocated to different subscribers. Following ISDN terminology, the GSM traffic channels are also designated as Bm channel (mobile B channel) or Lm channel (lower-rate mobile channel, with half the bit rate). A Bm channel is a TCH for the transmission of bit streams of either 13 kbit/s of digitally coded speech or of data streams at 14.5, 12, 6 or 3.6 kbit/s. Lm channels are TCH channels with less transmission bandwidth than Bm channels and transport speech signals of half the bit rate (TCH/H) or bit streams for data services with 6 or 3.6 kbit/s.

4.1.2   Signaling channels

The control and management of a cellular network demands a very high signaling effort. Even when there is no active connection, signaling information (for example, location update information) is permanently transmitted over the air interface. The GSM signaling channels offer a continuous, packet-oriented signaling service to MSs in order to enable them to send and receive messages at any time over the air interface to the BTS. Following ISDN terminology, the GSM signaling channels are also called Dm channels (mobile D channel). They are further divided into Broadcast Channel (BCH), Common Control Channel (CCCH) and Dedicated Control Channel (DCCH) (see Table 4.1).

The unidirectional BCHs are used by the BSS to broadcast the same information to all MSs in a cell. The group of BCHs consists of three channels.

- Broadcast Control Channel (BCCH): On this channel, a series of information elements is broadcast to the MSs which characterize the organization of the radio network, such as radio channel configurations (of the currently used cell as well as of the neighboring cells), synchronization information (frequencies as well as frame numbering) and registration identifiers (LAI, CI, BSIC). In particular, this includes information about the structural organization (formats) of the CCCH of the local BTS. The BCCH is broadcast on the first frequency assigned to the cell (the so-called BCCH carrier).

- Frequency Correction Channel (FCCH): On the FCCH, information about correction of the transmission frequency is broadcast to the MSs; see section 4.2.2 (frequency correction burst).

- Synchronization Channel (SCH): The SCH broadcasts information to identify a BTS, i.e. BSIC; see Chapter 3. The SCH also broadcasts data for the frame synchronization of a MS, i.e. Reduced Frame Number (RFN) of the TDMA frame; see section 4.3.1.

FCCH and SCH are only visible within protocol Layer 1, since they are only needed for the operation of the radio subsystem. There is no access to them from Layer 2. In spite of this fact, the SCH messages contain data which are needed by Layer 3 for the administration of radio resources. These two channels are always broadcast together with the BCCH.

The CCCH is a point-to-multipoint signaling channel to deal with access management functions. This includes the assignment of dedicated channels and paging to localize a MS. It comprises the following.

- Random Access Channel (RACH): The RACH is the uplink portion of the CCCH. It is accessed from the mobile stations in a cell without reservation in a competitive multiple-access mode using the principle of slotted Aloha (Bertsekas and Gallager, 1987), to ask for a dedicated signaling channel for exclusive use by one MS for one signaling transaction.

- Access Grant Channel (AGCH): The AGCH is the downlink part of the CCCH. It is used to assign an SDCCH or a TCH to a MS.

- Paging Channel (PCH): The PCH is also part of the downlink of the CCCH. It is used for paging to find specific MSs.

- Notification Channel (NCH): The NCH is used to inform MSs about incoming group and broadcast calls.

The last type of signaling channel, the DCCH is a bidirectional point-to-point signaling channel. An Associated Control Channel (ACCH) is also a dedicated control channel, but it is assigned only in connection with a TCH or an SDCCH. The group of Dedicated/Associated Control Channels (D/ACCH) comprises the following.

- Stand-alone Dedicated Control Channel (SDCCH): The SDCCH is a dedicated point-to-point signaling channel (DCCH) which is not tied to the existence of a TCH ('stand-alone'), i.e. it is used for signaling between a MS and the BSS when there is no active connection. The SDCCH is requested from the MS via the RACH and assigned via the AGCH. After the completion of the signaling transaction, the SDCCH is released and can be reassigned to another MS. Examples of signaling transactions which use an SDCCH are the updating of location information or parts of the connection setup until the connection is switched through (see Figure 4.1).

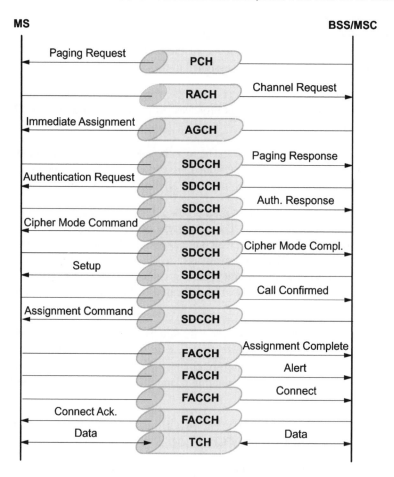

Figure 4.1 Logical channels and signaling (connection setup for an incoming call).

- Slow Associated Control Channel (SACCH): An SACCH is always assigned and used with a TCH or an SDCCH. The SACCH carries information for the optimal radio operation, e.g., commands for synchronization and transmitter power control and reports on channel measurements (section 4.5). Data must be transmitted continuously over the SACCH since the arrival of SACCH packets is taken as proof of the existence of the physical radio connection (section 4.5.3). When there is no signaling data to transmit, the MS sends a measurement report with the current results of the continuously conducted radio signal level measurements (section 4.5.1).

- Fast Associated Control Channel (FACCH): By using dynamic preemptive multiplexing on a TCH, additional bandwidth can be made available for signaling. The signaling channel created this way is called FACCH. It is only assigned in connection with a TCH, and its short-time usage goes at the expense of the user data transport.

Table 4.2 Logical channels of GSM protocol Layer 1.

Channel type	Net data throughput (kbit/s)	Block length (bits)	Block distance (ms)
TCH (full-rate speech)	13.0	$182 + 78$	20
TCH (half-rate speech)	5.6	$95 + 17$	20
TCH (data, 14.4 kbit/s)	14.5	290	20
TCH (data, 9.6 kbit/s)	12.0	60	5
TCH (data, 4.8 kbit/s)	6.0	60	10
TCH (data, up to 2.4 kbit/s)	3.6	72	10
FACCH full rate	9.2	184	20
FACCH half rate	4.6	184	40
SDCCH	598/765	184	3060/13
SACCH (with TCH)	115/300	$168 + 16$	480
SACCH (with SDCCH)	299/765	$168 + 16$	6120/13
BCCH	598/765	184	3060/13
AGCH	$n \times 598/765$	184	3060/13
NCH	$m \times 598/765$	184	3060/13
PCH	$p \times 598/765$	184	3060/13
RACH	$r \times 27/765$	8	3060/13
CBCH	598/765	184	3060/13

In addition to these channels, a Cell Broadcast Channel (CBCH) is defined, which is used to broadcast the messages of the Short Message Service Cell Broadcast (SMSCB). The CBCH shares a physical channel with the SDCCH.

4.1.3   Example: connection setup for incoming call

Figure 4.1 shows an example for an incoming call connection setup at the air interface. It is illustrated how the various logical channels are used in principle. The MS is called via the PCH and requests a signaling channel on the RACH. It obtains the SDCCH through an IMMEDIATE ASSIGNMENT message on the AGCH. Then follow authentication, start of ciphering and start of setup over the SDCCH. An ASSIGNMENT COMMAND message gives the traffic channel to the MS, which acknowledges its receipt on the FACCH of this traffic channel. The FACCH is also used to continue the connection setup.

4.1.4   Bit rates, block lengths and block distances

Table 4.2 gives an overview of the logical channels of Layer 1, the available bit rates, block lengths used and the intervals between transmission of blocks. The 14.4 kbit/s data service has been standardized in further GSM standardization phases. Note that the logical channels can suffer from substantial transmission delays depending on the respective use of forward error correction (channel coding and interleaving, see section 4.8 and Table 4.17 below).

Table 4.3 Channel combinations offered by the base station.

	B1	B2	B3	B4	B5	B6	B7	B8	B9
TCH/F	■							■	■
TCH/H		■	■						
TCH/H			■						
BCCH					■	■			
FCCH				■	■				
SCH				■	■				
CCCH				■	■				
SDCCH				■			■		
SACCH	■	■	■		■		■	■	
FACCH								■	

4.1.5   Combinations of logical channels

Not all logical channels can be used simultaneously at the radio interface. They can only be deployed in certain combinations and on certain physical channels. GSM has defined several channel configurations, which are realized and offered by the base stations (Table 4.3). As already mentioned before, an SACCH is always allocated either with a TCH or with an SDCCH, which accounts for the attribute 'associated'.

Depending on its current state, a MS can only use a subset of the logical channels offered by the base station. It uses the channels only in the combinations indicated in Table 4.4. The combination M1 is used in the phase when no physical connection exists, i.e. immediately after the power-up of the MS or after a disruption due to unsatisfactory radio signal conditions. Channel combinations M2 and M3 are used by active MSs in standby mode. In phases requiring a dedicated signaling channel, a MS uses the combination M4, whereas M5 to M8 are used when there is a traffic channel up. M8 is a multislot combination (a MS transmits on several physical channels), where n denotes the number of bidirectional channels, and m denotes the number of unidirectional channels ($n = 1, \ldots, 8, m = 0, \ldots, 7$, $n + m = 1, \ldots, 8$).

4.2   Physical channels

After discussing the logical channels and their tasks, we now deal with the physical channels, which transport the logical channels via the air interface. We first describe the GSM modulation technique (section 4.2.1), followed by the multiplexing structure (section 4.2.2): GSM is a multicarrier TDMA system, i.e. it employees a combination of FDMA and TDMA for multiple access. This section also covers the explanation of the radio bursts. Finally, section 4.2.3 briefly describes the (optional) frequency hopping technique, which has been standardized to reduce interference.

Table 4.4 Channel combinations used by the mobile station.

	M1	M2	M3	M4	M5	M6	M7	M8
TCH/F					▨			n+m
TCH/H						▨	▨	
TCH/H							▨	
BCCH	▨		▨					
CCCH		▨		▨				
SDCCH				▨				
SACCH					▨	▨	▨	n+m
FACCH					▨	▨	▨	

Figure 4.2 Steps involved in GSM digital modulation.

4.2.1 Modulation

The modulation technique used on the radio channel is Gaussian Minimum Shift Keying (GMSK). GMSK belongs to a family of continuous-phase modulation procedures, which have the special advantages of a narrow transmitter power spectrum with low adjacent channel interference, on the one hand, and a constant amplitude envelope, on the other hand, which allows use of simple amplifiers in the transmitters without special linearity requirements (class C amplifiers). Such amplifiers are especially inexpensive to manufacture, have high degree of efficiency and therefore allow longer operation on a battery charge (David and Benkner, 1996; Watson, 1993).

The digital modulation procedure for the GSM air interface comprises several steps for the generation of a high-frequency signal from channel-coded and enciphered data blocks (Figure 4.2).

The data d_i arrives at the modulator with a bit rate of $1625/6 = 270.83$ kbit/s (gross data rate) and are first differential-coded:

$$\hat{d}_i = (d_i + d_{i-1}) \bmod 2, \quad d_i \in (0; 1).$$

From this differential data, the modulation data are formed, which represents a sequence of Dirac pulses:

$$a_i = 1 - 2\hat{d}_i.$$

This bipolar sequence of modulation data is fed into the transmitter filter – also called a frequency filter – to generate the phase $\varphi(t)$ of the modulation signal. The impulse response $g(t)$ of this linear filter is defined by the convolution of the impulse response $h(t)$ of a

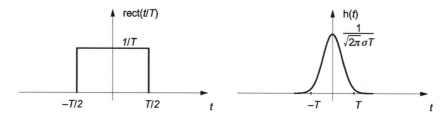

Figure 4.3 Impulse responses for the building blocks of the GMSK transmitter filter.

Gaussian low-pass with a rectangular step function:

$$g(t) = h(t) * \mathrm{rect}(t/T),$$

$$\mathrm{rect}(t/T) = \begin{cases} 1/T & \text{for } |t| < T/2, \\ 0 & \text{for } |t| \geq T/2, \end{cases}$$

$$h(t) = \frac{1}{\sqrt{2\pi}\,\sigma T}\, \exp\!\left(\frac{-t^2}{2\sigma^2 T^2}\right), \qquad \sigma = \frac{\sqrt{\ln 2}}{2\pi BT}, \qquad BT = 0.3.$$

In the equations above, B is the 3 dB bandwidth of the filter $h(t)$ and T the bit duration of the incoming bit stream. The rectangular step function and the impulse response of the Gaussian lowpass are shown in Figure 4.3, and the resulting impulse response $g(t)$ of the transmitter filter is given in Figure 4.4 for some values of BT. Note that with decreasing BT the impulse response becomes broader. For $BT \to \infty$ it converges to the rect() function.

In essence, this modulation consists of a Minimum Shift Keying (MSK) procedure, where the data are filtered through an additional Gaussian lowpass before Continuous Phase Modulation (CPM) with the rectangular filter (David and Benkner, 1996). Accordingly it is called GMSK. The Gaussian lowpass filtering has the effect of additional smoothing, but also of broadening the impulse response $g(t)$. This means that, on the one hand, the power spectrum of the signal is made narrower, but, on the other hand, the individual impulse responses are 'smeared' across several bit durations, which leads to increased intersymbol interference. This partial-response behavior has to be compensated for in the receiver by means of an equalizer (David and Benkner, 1996).

The phase of the modulation signal is the convolution of the impulse response $g(t)$ of the frequency filter with the Dirac impulse sequence a_i of the stream of modulation data:

$$\varphi(t) = \sum_i a_i \pi \eta \int_{-\infty}^{t-iT} g(u)\,\mathrm{d}u$$

with the modulation index at $\eta = 1/2$, i.e. the maximal phase shift is $\pi/2$ per bit duration. Accordingly, GSM modulation is designated as 0.3-GMSK with a $\pi/2$ phase shift. The phase $\varphi(t)$ is now fed to a phase modulator. The modulated high-frequency carrier signal can then be represented by the following expression, where E_c is the energy per bit of the modulated data rate, f_0 the carrier frequency and φ_0 is a random phase component staying constant during a burst:

$$x(t) = \sqrt{\frac{2E_c}{T}}\, \cos(2\pi f_0 t + \varphi(t) + \varphi_0).$$

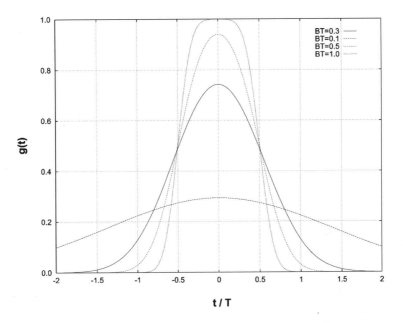

Figure 4.4 Impulse response $g(t)$ of the frequency filter (transmitter filter).

4.2.2 Multiple access, duplexing and bursts

At the physical layer (OSI Layer 1), GSM uses a combination of FDMA and TDMA for multiple access. Two frequency bands 45 MHz apart have been reserved for GSM operation (Figure 4.5): 890–915 MHz for transmission from the MS, i.e. uplink, and 935–960 MHz for transmission from the base station, i.e. downlink. Each of these bands of 25 MHz width is divided into 124 single carrier channels of 200 kHz width. This variant of FDMA is also called Multi-Carrier (MC). In each of the uplink/downlink bands there remains a guardband of 200 kHz. Each Radio Frequency Channel (RFCH) is uniquely numbered, and a pair of channels with the same number form a duplex channel with a duplex distance of 45 MHz (Figure 4.5).

A subset of the frequency channels, the Cell Allocation (CA), is allocated to a base station, i.e. to a cell. One of the frequency channels of the CA is used for broadcasting the synchronization data (FCCH and SCH) and the BCCH. Therefore, this channel is also called the BCCH carrier (see section 4.4). Another subset of the cell allocation is allocated to a MS, the Mobile Allocation (MA). The MA is used among others for the optional frequency hopping procedure (section 4.2.3). Countries or areas which allow more than one mobile network to operate in the same area of the spectrum must have a licensing agency which distributes the available frequency number space (e.g. the Federal Communication Commission in the USA or the 'Bundesnetzagentur' in Germany), in order to avoid collisions and to allow the network operators to perform independent network planning. Here is an example for a possible division: Operator A uses RFCH 2–13, 52–81 and 106–120, whereas operator B receives RFCH 15–50 and 83–103, in which case RFCH 1, 14, 51, 82, 104, 105 and 121–124 are left unused as additional guard bands.

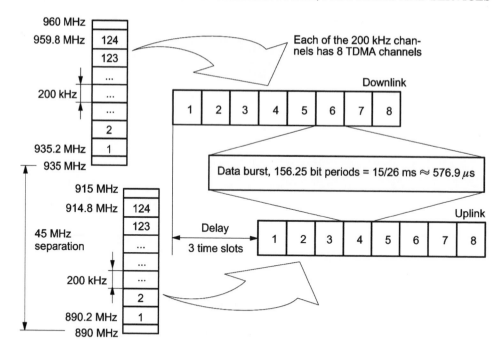

Figure 4.5 Carrier frequencies, duplexing and TDMA frames.

Each of the 200 kHz channels is divided into eight time slots and thus carries eight TDMA channels. The eight time slots together form a TDMA frame (Figure 4.5). The TDMA frames of the uplink are transmitted with a delay of three time slots with regard to the downlink (see Figure 4.6). A MS uses the same time slots in the uplink as in the downlink, i.e. the time slots with the same number (TN). Owing to the shift of three time slots, a MS does not have to send at the same time as it receives, and therefore does not need a duplex unit. This reduces the high-frequency requirements for the front end of the mobile and allows it to be manufactured as a less expensive and more compact unit.

So in addition to the separation into uplink and downlink bands – FDD with a distance of 45 MHz – the GSM access procedure contains a TDD component. Thus, the MS does not need its own high-frequency duplexing unit, which again reduces cost as well as energy consumption.

Each time slot of a TDMA frame lasts for a duration of 156.25 bit periods and, if used, contains a data burst. The time slot lasts $15/26$ ms $= 576.9$ μs; so a frame takes 4.615 ms. The same result is also obtained from the GMSK procedure, which realizes a gross data transmission rate of 270.83 kbit/s per carrier frequency.

There are five kinds of burst (Figure 4.7):

- Normal Burst (NB): The NB is used to transmit information on traffic and control (except RACH) channels. The individual bursts are separated from each other by guard periods during which no bits are transmitted. At the start and end of each burst are three tail bits which are always set to logical '0'. These bits fill a short time span

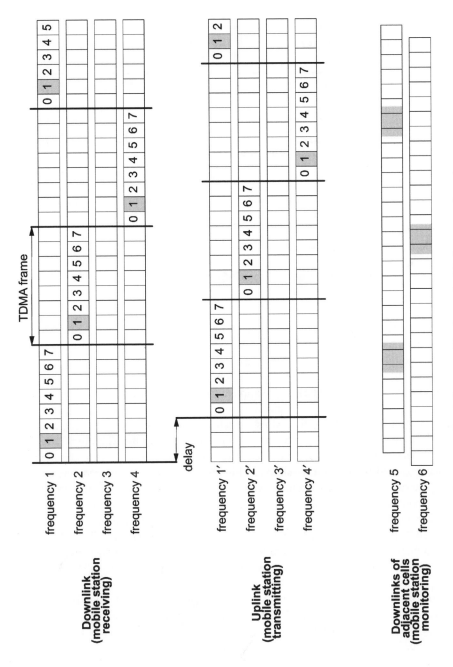

Figure 4.6 GSM full-rate traffic channel with frequency hopping.

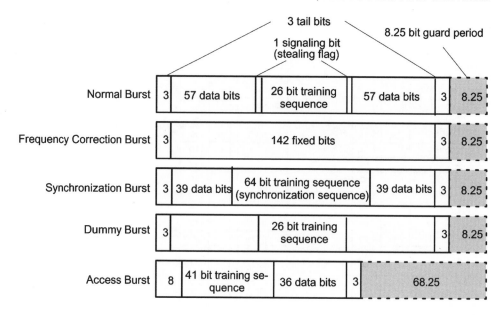

Figure 4.7 Bursts of the GSM TDMA procedure.

during which transmitter power is ramped up or ramped down and during which no data transmission is possible. Furthermore, the initial zero bits are also needed for the demodulation process. The Stealing Flags (SFs) are signaling bits which indicate whether the burst contains traffic data or signaling data. They are set to allow use of single time slots of the TCH in preemptive multiplexing mode, e.g. when, during a handover, fast transmission of signaling data on the FACCH is needed. This causes a loss of user data, i.e. these time slots are 'stolen' from the traffic channel, hence the name SF. In addition to the synchronization and signaling bits (Figure 4.7), the NB also contains two blocks of 57 bits each of error-protected and channel-coded user data separated by a 26-bit midamble. This midamble consists of predefined, known bit patterns, the training sequences, which are used for channel estimation to optimize reception with an equalizer and for synchronization. With the help of these training sequences, the equalizer eliminates or reduces the intersymbol interferences which are caused by propagation time differences of the multipath propagation. Time differences of up to 16 μs can be compensated for. Eight different training sequences are defined for the NB which are designated by the Training Sequence Code (TSC). Initially, the TSC is obtained when the BCC is obtained, which is transmitted as part of the BSIC (see Chapter 3). Beyond that, training sequences can be individually assigned to mobile stations. In this case the TSC is contained in the Layer 3 message of the channel assignment (TCH or SDCCH). In this way the base station tells a MS which training sequence it should use with NBs of a specific traffic channel.

- Frequency Correction Burst (FB): This burst is used for the frequency synchronization of a MS. The repeated transmission of FBs is also called the FCCH. Tail bits as well as data bits are all set to 0 in the FB. Owing to the GSM modulation procedure (0.3-GMSK) this corresponds to broadcasting an unmodulated carrier with a frequency shift of 1625/24 kHz above the nominal carrier frequency. This signal is periodically transmitted by the base station on the BCCH carrier. It allows time synchronization with the TDMA frame of a MS as well as the exact tuning to the carrier frequency. Depending on the stability of its own reference clock, the mobile can periodically resynchronize with the base station using the FCCH.

- Synchronization Burst (SB): This burst is used to transmit information which allows the MS to synchronize time-wise with the BTS. In addition to a long midamble, this burst contains the running number of the TDMA frame, the RFN and the BSIC; the RFN is covered in section 4.3. Repeated broadcasting of SBs is considered as the SCH.

- Dummy Burst (DB): This burst is transmitted on one frequency of the CA, when no other bursts are to be transmitted. The frequency channel used is the same as that which carries the BCCH, i.e. it is the BCCH carrier. This ensures that the BCCH transmits a burst in each time slot which enables the MS to perform signal power measurements of the BCCH, a procedure also known as quality monitoring.

- Access Burst (AB): This burst is used for random access to the RACH without reservation. It has a guard period significantly longer than the other bursts. This reduces the probability of collisions, since the MSs competing for the RACH are not (yet) time-synchronized.

A single user gets one-eighth or 33.9 kbit/s of the gross data rate of 270.83 kbit/s. Considering a normal burst, 9.2 kbit/s are used for signaling and synchronization, i.e. tail bits, SFs and training sequences, including guard periods. The remaining 24.7 kbit/s are available for the transmission of (raw) user or control data on the physical layer.

4.2.3 Optional frequency hopping

Mobile radio channels suffer from frequency-selective interferences, e.g. frequency-selective fading due to multipath propagation phenomena. This selective frequency interference can increase with the distance from the base station, especially at the cell boundaries and under unfavorable conditions. Frequency hopping procedures change the transmission frequencies periodically and thus average the interference over the frequencies in one cell. This leads to a further improvement in the Signal-to-Noise Ratio (SNR) to a high enough level for good speech quality, so that conversations with acceptable quality can be conducted. GSM systems achieve a good speech quality with a SNR of about 11 dB. With frequency hopping a value of 9 dB is sufficient. GSM provides for an optional frequency hopping procedure which changes to a different frequency with each burst; this is known as slow frequency hopping. The resulting hopping rate is about 217 changes per second, corresponding to the TDMA frame duration. The frequencies available for hopping, the hopping assignment, are taken from the CA. The principle is illustrated in Figure 4.6, showing the time slot allocations for a full-rate TCH. The exact synchronization is determined by several parameters: the MA, a Mobile Allocation Index Offset (MAIO), a Hopping Sequence Number (HSN) and the TDMA Frame Number (FN); see section 4.3. The use of frequency hopping is an option left to the network operator, which can be decided on an individual cell basis. Therefore, a MS must be able to switch to frequency hopping if a base station notices adverse conditions and decides to activate frequency hopping.

4.2.4 Summary

A physical GSM channel is defined by a sequence of frequencies and a sequence of TDMA frames. The RFCH sequence is defined by the frequency hopping parameters, and the temporal sequence of time slots of a physical channel is defined as a sequence of frame numbers and the time slot number within the frame. Frequencies for the uplink and downlink are always assigned as a pair of frequencies with a 45 MHz duplex separation.

As shown above, GSM uses a series of parameters to define a specific physical channel of a base station. Summarizing, these parameters are:

- MAIO;

- HSN;

- TSC;

- Time Slot Number (TN);

- MA, also known as RFCH Allocation;

- type of logical channel carried on this physical channel;

- the number of the logical subchannel (if used) – the Subchannel Number (SCN).

Within a logical channel, there can be several subchannels (e.g. subrate multiplexing of the same channel type). The TDMA frame sequence can be derived from the type of the channel and the logical subchannel if present.

4.3 Synchronization

For the successful operation of a mobile radio system, synchronization between MSs and the base station is necessary. Two kinds of synchronization are distinguished: frequency synchrony and time synchrony of the bits and frames.

Frequency synchronization is necessary so that transmitter and receiver frequencies agree. The objective is to compensate for tolerances of the less-expensive and, therefore, less-stable oscillators in the mobile stations by obtaining an exact reference from the base station and to follow it.

Bit and frame synchrony are important in two regards for TDMA systems. First, the propagation time differences of signals from different MSs have to be adjusted, so that the transmitted bursts are received synchronously with the time slots of the base station and that bursts in adjacent time slots do not overlap and interfere with each other. Second, synchrony is needed for the frame structure since there is a higher-level frame structure superimposed on the TDMA frames for multiplexing logical signaling channels onto one physical channel. The synchronization procedures defined for GSM are explained in the following section.

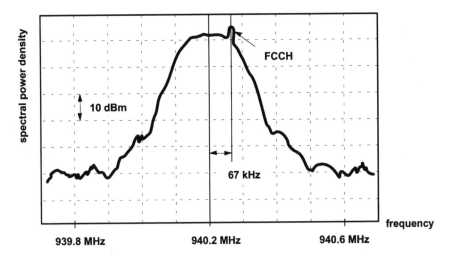

Figure 4.8 Typical power spectrum of a BCCH carrier.

4.3.1 Frequency and clock synchronization

A GSM base station transmits signals on the frequency carrier of the BCCH which allow a MS to synchronize with the base station. Synchronization means on the one hand the time-wise synchronization of the MS and base station with regards to bits and frames, and on the other hand tuning the MS to the correct transmitter and receiver frequencies.

For this purpose, the BTS provides the following signals (Figure 4.7):

- SCH with SBs and extra long training sequences, which facilitate synchronization;

- FCCH with FBs.

Owing to the 0.3-GMSK modulation procedure used in GSM, a data sequence of logical '0' generates a pure sine wave signal, i.e. broadcasting of the FB corresponds to an unmodulated carrier (frequency channel) with a frequency shift of 1625/24 kHz ($\approx$ 67.7 kHz) above the nominal carrier frequency (Figure 4.8). In this way, the MS can keep exactly synchronized by periodically monitoring the FCCH. On the other hand, if the frequency of the BCCH is still unknown, the MS can search for the channel with the highest signal level. This channel is with all likelihood a BCCH channel, because DBs must be transmitted on all unused time slots in this channel, whereas not all time slots are always used on other carrier frequencies. Using the FCCH sine wave signal allows identification of a BCCH and synchronization of a MSs oscillator.

For the time synchronization, TDMA frames in GSM are cyclically numbered *modulo* 2 715 648 ($= 26 \times 51 \times 2^{11}$) with the FN. One cycle generates the so-called hyperframe structure which comprises 2 715 648 TDMA frames. This long numbering cycle of TDMA frames is used to synchronize the ciphering algorithm at the air interface (see section 5.6). Each base station BTS periodically transmits the RFN on the SCH. With each SB the mobiles thus receive information about the number of the current TDMA frame. This enables each MS to be time-synchronized with the base station.

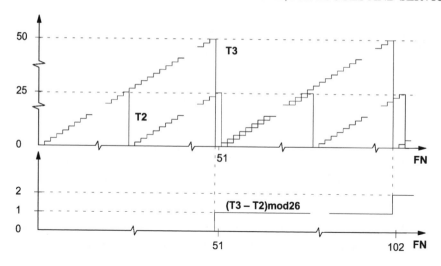

Figure 4.9 Values T2 and T3 for the calculation of RFN.

The RFN has a length of 19 bits. It consists of three fields: T1 (11 bits), T2 (5 bits) and T3′ (3 bits). These three fields are defined by (with div designating integer division)

$$T1 = FN \text{ div } (26 \times 51) \ [0\text{–}2047],$$

$$T2 = FN \text{ mod } 26 \ [0\text{–}25],$$

$$T3' = (T3 - 1) \text{ div } 10 \ [0\text{–}4], \quad \text{with } T3 = FN \text{ mod } 51 \ [0\text{–}50].$$

The sequences of running values of T2 and T3 are illustrated in Figure 4.9. The value crucial for the reconstruction of the FN is the difference between the two fields. The time synchronization of a MS and its time slots, TDMA frames and control channels is based on a set of counters which run continuously, independent of MS or base station transmission. Once these counters have been started and initialized correctly, the MS is in a synchronized state with the base station. The following four counters are kept for this purpose:

- quarter bit counter counting the Quarter Bit Number (QN);

- bit counter counting the Bit Number (BN);

- time slot counter counting the TN;

- frame counter counting the FN.

Owing to the bit and frame counting, these counters are of course interrelated, namely in such a way that the subsequent counter counts the overflows of the preceding counter. The following principle is used (Figure 4.10): QN is incremented every 12 or 13 μs; BN is obtained from this by integer division (BN = QN div 4). With each transition from 624 to 0 TN is incremented, and each overflow of TN increments the frame counter FN by 1.

The timers can be reset and restarted when receiving an SB. The quarter bit counter is set by using the timing of the training sequence of the burst, whereas the TN is reset to 0 with

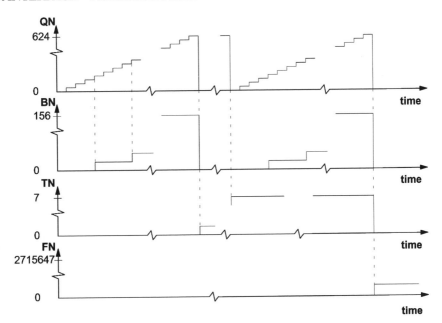

Figure 4.10 Synchronization timers, simplified: the TDMA frame duration is 156.25 bit times.

the end of the burst. The FN can then be calculated from the RFN transmitted on the SCH:

$$FN = 51 \times ((T3 - T2) \bmod 26) + T3 + 51 \times 26 \times T1, \quad \text{with } T3 = 10 \times T3' + 1.$$

It is important to recalculate T3 from T3′ although, because of the binary representation, only the integer part of the division by 10 is taken into account.

If the optional frequency hopping procedure is used (see section 4.2.3), an additional mapping of the TDMA frame number onto the frequency to be used is required in addition to the evaluation of the synchronization signals from the FCCH and SCH. One has to obtain the index number of the frequency channel on which the current burst has to be transmitted from the MA table. This process uses a predefined RFNTABLE, the FN and a HSN (Figure 4.11). The MA holds N frequencies, with a maximum value of 64 for N. With this procedure, every burst is sent on a different frequency in a cyclic way.

4.3.2 Adaptive frame synchronization

The MS can be anywhere within a cell, which means the distance between MS and base station may vary. Thus, the signal propagation times between MS and base station vary. Owing to the mobility of the subscribers, the bursts received at the base would be offset. The TDMA procedure cannot tolerate such time shifts, since it is based on the exact synchronization of transmitted and received data bursts. Bursts transmitted by different MSs in adjacent time slots must not overlap when received at the base station by more than the guard period (Figure 4.7), even if the propagation times within the cell are very different.

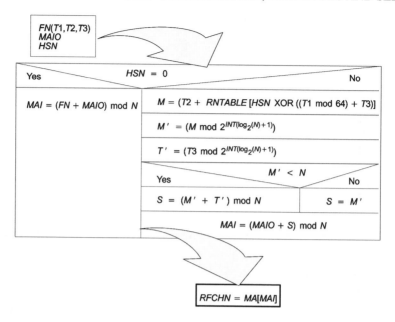

Figure 4.11 Generation of the GSM frequency hopping sequence.

To avoid such collisions, the start of transmission time from the MS is advanced in proportion to the distance from the base station. The process of adapting the transmissions from the MSs to the TDMA frame is called adaptive frame alignment.

For this purpose, the parameter Timing Advance (TA) in each SACCH Layer 1 protocol block is used (section 4.5). The MS receives the TA value it must use from the base station on the SACCH downlink; it reports the actually used value on the SACCH uplink. There are 64 steps for the timing advance which are coded as 0 to 63. One step corresponds to one bit period. Step 0 means no timing advance, i.e. the frames are transmitted with a time shift of three slots or 468.75 bit durations with regards to the downlink. At step 63, the timing of the uplink is shifted by 63 bit durations, such that the TDMA frames are transmitted on the uplink only with a delay of 405.75 bit durations. So the required adjustment always corresponds to twice the propagation time or is equal to the round-trip delay (Figure 4.12). In this way, the available range of values allows a compensation over a maximum propagation time of 31.5 bit periods ($\approx$113.3 μs). This corresponds to a maximum distance between mobile and base station of 35 km. A GSM cell may therefore have a maximum diameter of 70 km. The distance from the base station or the currently valid TA value for a MS is therefore an important handover criterion in GSM networks (section 6.4.3).

The adaptive frame alignment technique is based on continuous measurement of propagation delays by the base station and corresponding timing advance activity by the MS. In the case of an (unreserved) random access to the RACH, a channel must first be established. The base station has in this case not yet had the opportunity to measure the distance to the MS and to transmit a corresponding timing advance command. If a MS transmits an access burst in the current time slot, it uses a timing advance value of 0 or a default value. To minimize collisions

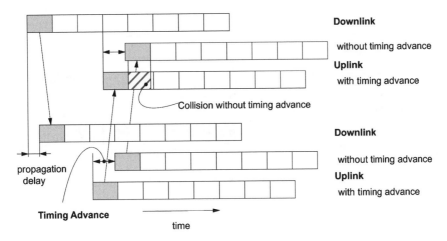

Figure 4.12 Operation of timing advance.

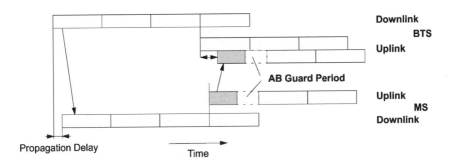

Figure 4.13 Timing for RACH random multiple access.

with subsequent time slots at the base station, the access burst (AB) has to be correspondingly shorter than the time slot duration (Figure 4.13). This explains the long guard period of the AB of 68.25 bit periods, which can compensate for the propagation delay if a MS sends an AB from the boundary of a cell of 70 km diameter.

4.4 Mapping of logical onto physical channels

The mapping of logical channels onto physical channels has two components: mapping in frequency and mapping in time. The mapping of a logical channel onto a physical channel in the frequency domain is based on the FN, the frequencies allocated to base stations and MSs – CA and MA – and the rules for the optional frequency hopping (see section 4.2.3).

In the time domain, logical channels are transported in the corresponding time slots of the physical channel. They are mapped onto physical channels in certain time-multiplexed combinations, where they can occupy a complete physical channel or just part of a physical

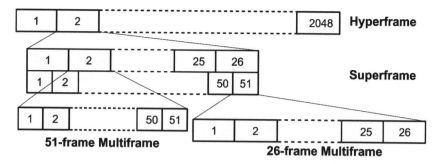

Figure 4.14 GSM frame structures.

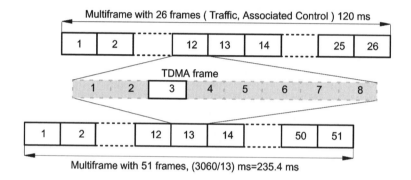

Figure 4.15 GSM multiframes.

channel. Whereas user payload data is allocated a dedicated full-rate or half-rate channel, logical signaling (control) channels have to share a physical channel.

The logical channels are organized by the definition of complex superstructures on top of the TDMA frames, forming so-called multiframes, superframes and hyperframes (Figure 4.14). For the mapping of logical onto physical channels, we are interested in the multiframe domain. These multiframes allow us to map (logical) subchannels onto physical channels. Two kinds of multiframes are defined (Figure 4.15): a multiframe consisting of 26 TDMA frames (predominantly payload – speech and data – frames) and a multiframe of 51 TDMA frames (predominantly signaling frames).

Each hyperframe is divided into 2048 superframes. With its long cycle period of 3 h 28 min 53.760 s, it is used for the synchronization of user data encryption. A superframe consists of 1326 consecutive TDMA frames which therefore lasts for 6.12 s, the same length as 51 multiframes of 26 TDMA frames or 26 multiframes of 51 TDMA frames. These multiframes are again used to multiplex the different logical channels onto a physical channel as shown below.

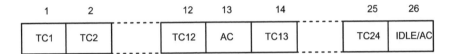

Figure 4.16 Channel organization in a 26-frame multiframe.

4.4.1   26-frame multiframe

Each 26 subsequent TDMA frames form a multiframe which multiplexes two logical channels, a TCH and the SACCH, onto the physical channel (Figure 4.16). This process uses only one time slot per TDMA frame for the corresponding multiframe (e.g. time slot 3 in Figure 4.15), since a physical channel consists of just one time slot per TDMA frame. In addition to the 24 TCH frames for user data, this multiframe also contains an Associated Control (AC) frame for signaling data (SACCH data). One frame (the 26th) remains unused in the case of a full-rate TCH (IDLE/AC); it is reserved for the introduction of two half-rate TCHs; then the 26th frame will be used to carry the SACCH channels of the other half.

The data of the FACCH is transmitted by occupying one-half of the bits in eight consecutive bursts, by 'stealing' these bits from the TCH. For this purpose, the SFs of the normal bursts are set (Figure 4.7).

A subscriber has available a gross data rate of $271/8 = 33.9$ kbit/s (section 4.2). Of this budget, 9.2 kbit/s are for signaling, synchronization and guard periods of the burst. Of the remaining 24.7 kbit/s, in the case of the 26-frame multiframe, 22.8 kbit/s are left for the coded and enciphered user data of a full-rate channel and 1.9 kbit/s remain for the SACCH and IDLE.

4.4.2   51-frame multiframe

For the transmission of the control channels which are not associated with a TCH (all except FACCH and SACCH), a multiframe is formed consisting of 51 consecutive TDMA frames (Figure 4.7). According to the channel configuration (section 4.1), the multiframe is used differently. In each case, multiframes of 51 TDMA frames serve the purpose of mapping several logical channels onto a physical channel.

Furthermore, some of these control channels are unidirectional, which results in different structures for uplink and downlink. For some configurations, two adjacent multiframes are required to map all of the logical channels. Some examples are illustrated in Figure 4.17. They correspond to the combinations B2, B3 and B4 in Table 4.3 whereas for channels SDCCH and SACCH some four or eight logical subchannels have been defined (D0, D1, . . . , A0, A1, . . .). One of the frequency channels of the CA of a base station is used to broadcast synchronization data (FCCH and SCH) and the BCCH. Since the base station has to transmit in each time slot of the BCCH carrier to enable a continuous measurement of the BCCH carrier by the MS, a DB is transmitted in all time slots with no traffic.

On time slot 0 of the BCCH carrier, only two combinations of logical channels may be transmitted, the combinations B2 or B3 from Table 4.3: (BCCH + CCCH + FCCH + SCH + SDCCH + SACCH or BCCH + CCCH + FCCH + SCH). No other time slot of the CA must carry this combination of logical channels.

Table 4.5   Mapping of the frame number onto a BCCH message.

TC	System information message
0	Type 1
1	Type 2
2, 6	Type 3
3, 7	Type 4
4, 5	Any (optional)

As one can see in Figure 4.17, in the time slot 0 of the BCCH carrier of a base station (downlink) the frames 1, 11, 21, . . . are FCCH frames, and the subsequent frames 2, 12, 22, . . . form SCH frames. Frames 3, 4, 5 and 6 of the 51-frame BCCH multiframe transport the appropriate BCCH information, whereas the remaining frames may contain different combinations of logical channels. Once the MS has synchronized by using the information from FCCH and SCH, it can determine from the information in the FCCH and SCCH how the remainder of the BCCH is constructed. For this purpose, the base station radio resource management periodically transmits a set of messages to all MSs in this cell.

These system information messages comprise six types, of which only types 1–4 are of interest here. Using FN, one can determine which type is to be sent in the current time slot by calculating a Type Code (TC):

$$TC = (FN \text{ div } 51) \bmod 8.$$

Table 4.5 shows how the TC determines the type of the system information message to be sent within the current multiframe.

Of the parameters contained in such a message, the following are of special interest. BS_CC_CHANS determines the number of physical channels which support a CCCH. The first CCCH is transmitted in time slot 0, the second in time slot 2, the third in time slot 4 and the fourth in time slot 6 of the BCCH carrier. Another parameter, BS_CCCH_SDCCH_COMB, determines whether the DCCHs SDCCH(0-3) and SACCH(0-3) are transmitted together with the CCCH on the same physical channel. In this case, each of these dedicated control channels consists of four subchannels.

Each of the CCCHs of a base station is assigned a group CCCH_GROUP of MSs. MSs are allowed random access (RACH) or receive paging information (PCH) only on the CCCH assigned to this group. Furthermore, a MS needs only to listen for paging information on every Nth block of the PCH. The number N is determined by multiplying the number of paging blocks per 51-frame multiframe of a CCCH with the parameter BS_PA_MFRMS designating the number of multiframes between paging frames of the same paging group (PAGING_GROUP).

In cells with high traffic, in particular, the CCCH and paging groups serve to subdivide traffic and to reduce the load on the individual CCCHs. For this purpose, there is a simple algorithm which allows each MS to calculate its respective CCCH_GROUP and PAGING_GROUP from its IMSI and parameters BS_CC_CHANS, BS_PA_MFRMS and N.

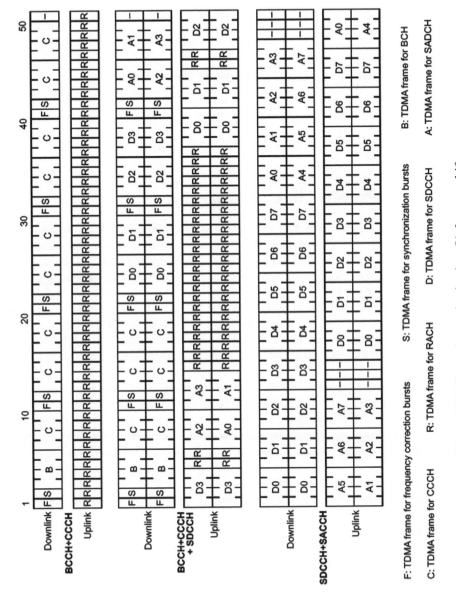

Figure 4.17 Channel organization in a 51-frame multiframe.

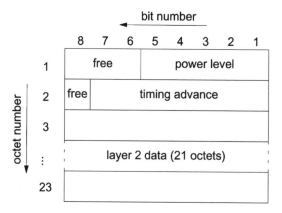

Figure 4.18 SACCH block format.

4.5   Radio subsystem link control

The radio interface is characterized by another set of functions of which we discuss only the most important in the following. One of these functions is the control of the radio link: radio subsystem link control, with the main activities of received-signal quality measurement (quality monitoring) for cell selection and handover preparation, and of transmitter power control.

If there is no active connection, i.e. if the MS is at rest, the BSS has no tasks to perform. The MS, however, is still committed to continuously observing the BCCH carrier of the current and neighboring cells, so that it would be able to select the cell in which it can communicate with the highest probability. If a new cell needs to be selected, a location update may become necessary.

During a connection (TCH or SDCCH), the functions of channel measurement and power control serve to maintain and optimize the radio channel; this also includes adaptive frame alignment (section 4.3.1) and frequency hopping (section 4.2.3). Both need to be done until the current base can hand over the current connection to the next base station.

These link control functions are performed over the SACCH channel. Two fields are defined in an SACCH block (Figure 4.18) for this purpose, the power level and the TA. On the downlink, these fields contain values as assigned by the BSS. On the uplink, the MS inserts its currently used values. The quality monitoring measurement values are transmitted in the data part of the SACCH block.

The following illustrates the basic operation of the radio subsystem link control at the BSS side for an existing connection; the detailed explanation of the respective functions is given later. In principle, the radio link control can be subdivided into three tasks: measurement collection and processing, transmitter power control, and handover control.

In the example of Figure 4.19, the process BSS_Link_Control starts at initialization the processes BSS_Power_Control and BSS_HO_Control and then enters a measurement loop, which is only left when the connection is terminated. In this loop, measurement data are periodically received (every 480 ms) and current mean values are calculated. At first, these measurement data are supplied to the transmitter power control to adapt the power of MS and

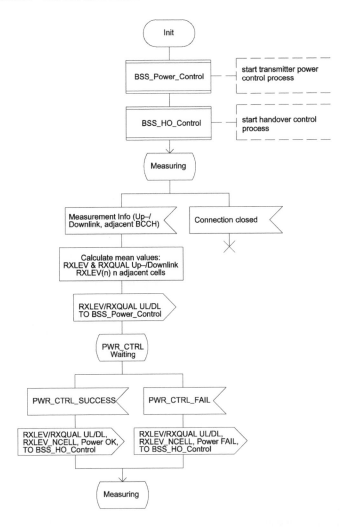

Figure 4.19 Principal operation of the radio subsystem link control.

BSS to a new situation if necessary. Thereafter, the measurement data and the result of the power control activity are supplied to the handover process, which can then decide whether a handover is necessary or not.

4.5.1 Channel measurement

The task of radio subsystem link control in the MS includes identification of the reachable base stations and measurement of their respective received signal level and channel quality (quality monitoring task). In idle mode, these measurements serve to select the current base

Table 4.6 Measurement range of the received signal level.

Level	Received signal level (dBm)	
	From	To
RXLEV_0	–	−110
RXLEV_1	−110	−109
⋮	⋮	⋮
RXLEV_62	−49	−48
RXLEV_63	−48	–

station, whose PCH is then periodically examined and on whose RACH desired connections can be requested.

During a connection, i.e. on a TCH or SDCCH with respective SACCH/FACCH, this measurement data are transmitted on the SACCH to the base station as a measurement report/measurement info. These reports serve as inputs for the handover and power control algorithms.

The measurement objects are, on the one hand, the uplink and downlink of the current channel (TCH or SDCCH) and, on the other hand, the BCCH carriers which are continuously broadcast with constant power by all BTSs in all time slots. It is especially important to keep the transmitter power of the BCCH carriers constant to allow comparisons between neighboring base stations. A list of neighboring base station's BCCH carrier frequencies, called the BCCH Allocation (BA) is supplied to each mobile by its current BTS, to enable measurement of all cells which are candidates for a handover. The cell identity is broadcast as the BSIC on the BCCH. Furthermore, up to 36 BCCH carrier frequencies and their BSICs can be stored on the SIM card. In principle, GSM uses two parameters to describe the quality of a channel: the Received Signal Level (RXLEV), measured in dBm, and the Received Signal Quality (RXQUAL), measured as bit error ratio as a percentage before error correction (Tables 4.6 and 4.7). The received signal power is measured continuously by MSs and base stations in each received burst within a range of -110 to -48 dBm. The respective RXLEV values are obtained by averaging.

The bit error ratio before error correction can be determined in a variety of ways. For example, it can be estimated from information obtained from channel estimation for equalization from the training sequences, or the number of erroneous (corrected) bits can be determined through repeated coding of the decoded, error-corrected data blocks and comparison with the received data. Since the data before error correction is presented as blocks of 456 bits (section 4.8), the bit error ratio can only be given with a quantizing resolution of 2×10^{-3}. Again, the value of RXQUAL is determined from this information by averaging.

Channel measurement during idle mode

In idle mode (see also Figure 5.17) the MS must always stay aware of its environment. The main purpose is to be able to assign a MS to a cell, whose BCCH carrier it can decode reliably. If this is the case, the MS is able to read system and paging information. If there is a

Table 4.7 Measurement range of the bit error ratio.

	Bit error ratio (%)	
Level	From	To
RXQUAL_0	–	0.2
RXQUAL_1	0.2	0.4
RXQUAL_2	0.4	0.8
RXQUAL_3	0.8	1.6
RXQUAL_4	1.6	3.2
RXQUAL_5	3.2	6.4
RXQUAL_6	6.4	12.8
RXQUAL_7	12.8	–

desire to set up a connection, the MS can most likely communicate with the network. There are two possible starting situations:

- the MS has no *a priori* knowledge about the network at hand, especially which BCCH carrier frequencies are in use;

- the MS has a stored list of BCCH carriers.

In the first case, the more unfavorable of the two, the mobile has to search through all of the 124 GSM frequencies, measure their signal power level and calculate an average from at least five measurements. The measurements of the individual carriers should be evenly distributed over an interval of 3–5 s. After at most 5 s, a minimum of 629 measurement values are available that allow the 124 RXLEV values to be determined. The carriers with the highest RXLEV values are very likely BCCH carriers, since continuous transmission is required on them. Final identification occurs with the FB of the FCCH. Once the received BCCH carriers have been found, the MS starts to synchronize with each of them and reads the system information, beginning with the BCCH with the highest RXLEV value.

This orientation concerning the current location can be accelerated considerably, if a list of BCCH carriers has been stored on the SIM card. Then the MS tries first to synchronize with some known carrier. Only if it cannot find any of the stored BCCH carrier frequencies, it does start with the normal BCCH search. A MS can store several lists for the recently visited networks.

Channel measurement during a connection

During a traffic (TCH) or signaling (SDCCH) connection, the channel measurement of the MS occurs over an SACCH interval, which comprises 104 TDMA frames in the case of a TCH channel (480 ms) or 102 TDMA frames (470.8 ms) in the case of an SDCCH channel.

For the channel at hand, two parameters are determined: the received signal level RXLEV and the signal quality RXQUAL. These two values are averaged over an SACCH interval (480 or 470.8 ms) and transmitted to the base station on the SACCH as a measurement report/measurement info. In this way the downlink quality of the channel assigned to the MS

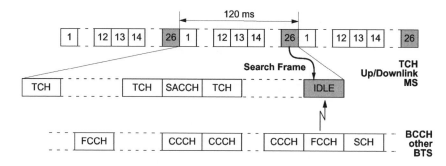

Figure 4.20 Synchronization with adjacent cells during a call.

can be judged. In addition to these measurements of the downlink by the MS, the base station also measures the RXLEV and RXQUAL values of the respective uplink.

In order to make a handover decision, information about possible handover targets must be available. For this purpose, the MS has to observe continuously the BCCH carriers of up to six neighboring base stations. The RXLEV measurements of the neighboring BCCH carriers are performed during the MSs unused time slots (see Figure 4.6). The BCCH measurement results of the six strongest signals are included in the measurement report transmitted to the BSS.

However, the received signal power level and the frequency of a BCCH carrier alone are not a sufficient criterion for a successful handover. Owing to the frequency reuse in cellular networks, and especially in the case of small clusters, it is possible that a cell can receive the same BCCH carrier from more than one neighboring cell, i.e. there exist several neighboring cells which use the same BCCH carrier. It is therefore necessary, to also know the identity (BSIC) of each neighboring cell. Simultaneously with the signal level measurement, the MS has to synchronize with each of the six neighboring BCCHs and read at least the SCH information.

For this purpose, one must first search for the FCCH burst of the BCCH carrier; then the SCH can be found in the next TDMA frame. Since the FCCH/SCH/BCCH is always transmitted in time slot 0 of the BCCH carrier, the search during a conversation for FCCHs can only be conducted in unused frames, i.e. in the case of a full-rate TCH in the IDLE frame of the multiframe (frame number 26 in Figures 4.16 and 4.20).

These free frames are therefore also known as search frames. Therefore, there are exactly four search frames within an SACCH block of 480 ms (four 26-frame multiframes of 120 ms). The MS has to examine the surrounding BCCH carriers for FCCH bursts, in order to synchronize with them and to decode the SCH. However, how can one search for synchronization points exactly within these frames during synchronized operation?

This is possible because the actual traffic channel and the respective BCCH carriers use different multiframe formats. Whereas the traffic channel uses the 26-frame multiframe format, time slot 0 of the BCCH carrier with the FCCH/SCH/BCCH is carried on a 51-frame multiframe format. This ratio of the different multiframe formats has the effect that the relative position of the search frames (frame 26 in a TCH multiframe) is shifting with regard to the BCCH multiframe by exactly one frame each 240 ms (Figure 4.21).

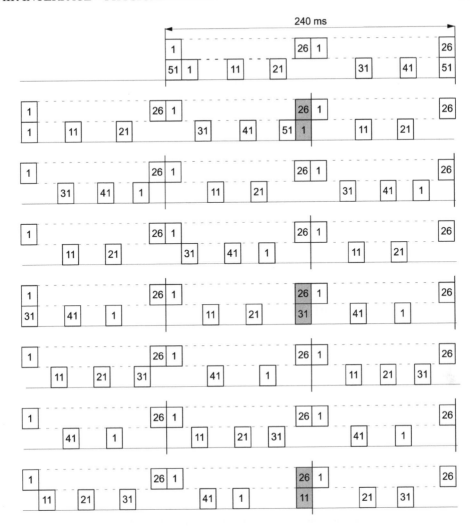

Figure 4.21 Principle of FCCH search during the search frame.

Figuratively speaking, the search frame is travelling along the BCCH multiframe in such a way that at most after 11 TCH multiframes (= 1320 ms) a frequency correction burst of a neighboring cell becomes visible in a search frame.

In this way, the MS is able to determine the BSIC for the respective RXLEV measurement value. Only BCCH carrier measurements whose identity can be established without doubt are included in the measurement report to the base station.

The base station can now make a handover decision based on these values, on the distance of the MS, and on the momentary interference of unused time slots.

The algorithm for handover decisions has not been included in the GSM standard. The network operators may use algorithms which are optimized for their network or the local

Table 4.8 GSM power classes.

Power class	Maximum peak transmission power (W)	
	Mobile station (dBm)	Base station
1	20 (43)	320
2	<8 (39)	160
3	<5 (37)	80
4	<2 (33)	40
5	<0.8 (29)	20
6	–	10
7	–	5
8	–	2.5

situation. GSM only gives a basic proposal which satisfies the minimum requirements for a handover decision algorithm. This algorithm defines threshold values, which must be violated in one or the other direction to arrive at a safe handover decision and to avoid so-called ping-pong handovers, which oscillate between two cells. Although the decision algorithm is part of radio subsystem link control, its discussion is postponed and it is treated together with handover signaling (see section 6.4.3).

4.5.2 Transmission power control

Power classes (Table 4.8) are used for classification of base stations and MSs. The transmission power can also be controlled adaptively. As part of the radio subsystem link control, the MS's transmitter power is controlled in steps of 2 dBm.

The GSM transmitter power control has the purpose of limiting the MS's transmitter power to the minimum necessary level, in such a way that the base station receives signals from different MSs at approximately the same power level. Sixteen power control steps are defined for this purpose: step 0 (43 dBm = 20 W) to step 15 (13 dBm). Starting with the lowest, step 15, the base station can increment the transmitter power of the MS in steps of 2 dBm up to the maximum power level of the respective power class of the MS. Similarly, the transmitter power of the base station can be controlled in steps of 2 dBm, with the exception of the BCCH carrier of the base station, which must remain constant to allow comparative measurements of neighboring BCCH carriers by the MSs.

Transmission power control is based on the measurement values RXLEV and RXQUAL, for which one has defined upper and lower thresholds for uplink and downlink (Table 4.9). Network management defines the adjustable parameters P and N. If the values of P for the last N calculated mean values of the respective criterion (RXLEV or RXQUAL) are above or below the respective threshold value, the BSS can adjust the transmitter power (Figure 4.22).

If the thresholds U_xx_UL_P of the uplink are exceeded, the transmission power of the MS is reduced; in the other case, if the signal level is below the threshold L_xx_UL_P, the MS is ordered to increase its transmitter power. In an analogous way, the transmitter power of the base station can be adjusted, when the criteria for the downlink are exceeded in either direction.

Table 4.9 Thresholds for transmitter power control.

Threshold parameter	Typical value (dBm)	Meaning
L_RXLEV_UL_P	−103 to −73	Threshold for raising of transmission power
L_RXLEV_DL_P	−103 to −73	in uplink or downlink
L_RXQUAL_UL_P	–	
L_RXQUAL_DL_P	–	
U_RXLEV_UL_P	–	Threshold for reducing of transmission
U_RXLEV_DL_P	–	power in uplink or downlink
U_RXQUAL_UL_P	–	
U_RXQUAL_DL_P	–	

Even if the MS or base station signal levels stay within the thresholds, the current RXLEV/RXQUAL values can cause a change to another channel of the same or another cell based on the handover thresholds (Table 6.1). For this reason, checking for transmitter thresholds is immediately followed by a check of the handover thresholds as the second part of the radio subsystem link control (Figures 4.19 and 6.17). If one of the threshold values is exceeded in either direction and the transmitter power cannot be adjusted accordingly, i.e. the respective transmitter power has reached its maximum or minimum value, this is an overriding cause for handover (PWR_CTRL_FAIL; Table 6.2) which the BSS must communicate immediately to the MSC (section 6.4).

4.5.3   Disconnection due to radio channel failure

The quality of a radio channel can vary considerably during an existing connection, or it can even fail in the case of shadowing. This should not lead to immediate disconnection, since such failures are often of short duration. For this reason GSM has a special algorithm within the Radio Subsystem Link Control which continuously checks for connectivity. It consists of recognizing a radio link failure by the inability to decode signaling information on the SACCH. This connectivity check is done both in the MS as well as in the base station. The connection is not immediately terminated, but is delayed so that only repeated consecutive failures (erroneous messages) represent a valid disconnect criterion. On the downlink, the MS must check the frequency of erroneous, nondecodable messages on the SACCH. The error protection on the SACCH has very powerful error correction capabilities and thus guarantees a very low probability of 10^{-10} for nonrecognized, wrongly corrected bits in SACCH messages.

In this way, erroneous SACCH messages supply a measure for the quality of the downlink, which is already quite low when errors on the SACCH cannot be corrected any more. If a consecutive number of SACCH messages is erroneous, the link is considered bad and the connection is terminated. For this purpose, a counter S has been defined which is incremented by 2 with each arrival of an error-free message, and decremented by 1 for each erroneous SACCH message (Figure 4.23). When the counter reaches the value $S = 0$, the downlink is considered as failing, and the connection is terminated. This failure is signaled to the

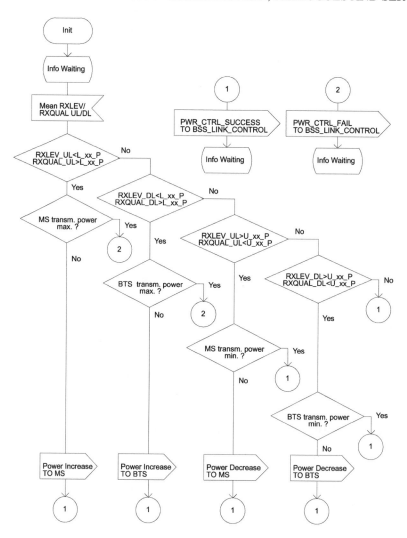

Figure 4.22 Schematic operation of transmitter power control.

upper layers, Mobility Management (MM), which can start a call reestablishment procedure. The maximum value RADIO_LINK_TIMEOUT for the counter S therefore determines the interval length during which a channel has to fail before a connection is terminated. After assignment of a dedicated channel (TCH or SDCCH), the MS starts the checking process by initializing the counter S with this value (Figure 4.23), which can be set individually per cell and is broadcast on the BCCH.

The corresponding checks are also conducted on the uplink. In both cases, however, this requires continuous transmission of data on the SACCH, i.e. when no signaling data have to be sent, filling data are transmitted. On the uplink, current measurement reports

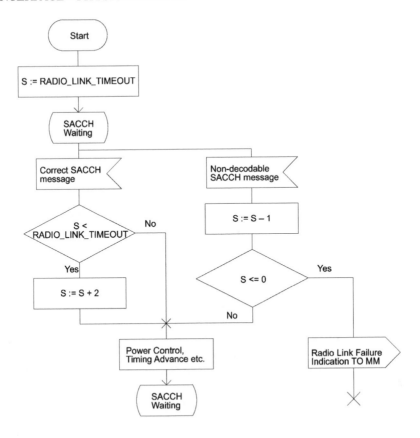

Figure 4.23 MS disconnect procedure.

are transmitted, whereas the downlink carries system information of Type 5 and Type 6 (section 5.4.3).

4.5.4 Cell selection and operation in power conservation mode

Cell selection and cell reselection

A MS in idle mode must periodically measure the receivable BCCH carriers of the base stations in the area and calculate mean values RXLEV(n) from this data (section 4.5.1). Based on these measurements, the MS selects a cell, namely that with the best reception, i.e. the MS is committed to this cell. This is called 'camping' on this cell. In this state, accessing a service becomes possible, and the MS listens periodically to the PCH. Two criteria are defined for the automatic selection of cells: the path loss criterion C1 and the reselection criterion C2. The path loss criterion serves to identify cell candidates for camping. For such cells, C1 has to be greater than zero. At least every 5 s, a MS has to recalculate C1 and C2 for the current and neighboring cells. If the path loss criterion of the current cell falls below zero, the path loss to the current base station has become too large. A new cell has to be selected, which requires use of the criterion C2. If one of the neighboring cells has a value of C2 greater than zero, it becomes the new current cell. The cell selection algorithm uses two further threshold values, which are broadcast on the BCCH:

- the minimum received power level RXLEV_ACCESS_MIN (typically −98 to −106 dBm) required for registration into the network of the current cell;

- the maximum allowed transmitter power MS_TXPWR_CCH (typically 31–39 dBm) allowed for transmission on a control channel (RACH) before having received the first power control command.

In consideration of the maximal transmitter power P of a mobile station, the path loss criterion C1 is now defined using the minimal threshold RXLEV_ACCESS_MIN for network access and the maximal allowed transmitter power MS_TXPWR_MAX_CCH:

$$C1(n) = (\text{RXLEV}(n) - \text{RXLEV\_ACCESS\_MIN}$$

$$- \text{maximum}(0, (\text{MS\_TXPWR\_MAX\_CCH} - P))).$$

The values of the path loss criterion C1 are determined for each cell for which a value RXLEV(n) of a BCCH carrier can be obtained. The cell with the lowest path loss can thus be determined using this criterion. It is the cell for which $C1 > 0$ has the largest value. During cell selection, the MS is not allowed to enter power conservation mode (Discontinuous Transmission (DTX), section 4.5.4).

A prerequisite for cell selection is that the cell considered belongs to the home Public Land Mobile Network (PLMN) of the MS or that access to the PLMN of this cell is allowed. Beyond that, a limited service mode has been defined with restricted service access, which still allows emergency calls if nothing else. In limited service mode, a MS can be camping on any cell but can only make emergency calls. Limited service mode exists when there is no SIM card in the MS, when the IMSI is unknown in the network or the IMEI is barred from service, but also if the cell with the best value of C1 does not belong to an allowed PLMN.

Once a MS is camping on a cell and is in idle mode, it should keep observing all of the BCCH carriers whose frequencies, the BA, are broadcast on the current BCCH. Having left idle mode, e.g. if a TCH has been assigned, the MS monitors only the six strongest neighboring BCCH carriers. A list of these six strongest neighboring BCCH carriers has already been prepared and kept up to date in idle mode. The BCCH of the camped-on cell must be decoded at least every 30 s. At least once every 5 min, the complete set of data from the six strongest neighboring BCCH carriers has to be decoded, and the BSIC of each of these carriers has to be checked every 30 s. This allows the MS to stay aware of changes in its environment and to react appropriately. In the worst case, conditions have changed so much that a new cell to camp on needs to be selected (cell reselection).

For this cell reselection, a further criterion C2, the reselection criterion, has been defined:

$$C2(n) = C1(n) + \text{CELL\_RESELECT\_OFFSET}$$

$$- (\text{TEMPORARY\_OFFSET} \times H(\text{PENALTY\_TIME} - T)),$$

$$\text{with } H(x) = \begin{cases} 0 & \text{for } x < 0, \\ 1 & \text{for } x \geq 0. \end{cases}$$

The interval T in this criterion is the time passed since the MS observed the cell n for the first time with a value of C1 > 0. It is set back to 0 when the path loss criterion C1 falls to C1 < 0. The parameters CELL_RESELECT_OFFSET, TEMPORARY_OFFSET, and PENALTY_TIME are announced on the BCCH. However, as a default, they are set to 0. Otherwise, the criterion C2 introduces a time hysteresis for cell reselection. It tries to ensure that the MS is camping on the cell with the highest probability of successful communication.

One exception for cell reselection is the case when a new cell belongs to another location area. In this case C2 must not only be larger than zero, but C2 > CELL_RESELECT_ HYSTERESIS to avoid too frequent location updates.

Discontinuous reception

To limit power consumption in idle mode and thus increase battery life in standby mode, the MS can activate the Discontinuous Reception (DRX) mode. In this mode, the receiver is turned on only for the phases of receiving paging messages and is otherwise in the power conservation mode which still maintains synchronization with BCCH signals through internal timers. In this DRX mode, measurement of BCCH carriers is performed only during unused time slots of the paging blocks.

4.6   Channel coding, source coding and speech processing

The previous sections explained the basic functions of the physical layer at the air interface, e.g. the definition of logical and physical channels, modulation, multiple access techniques, duplexing, and the definition of bursts. The following sections discuss several additional functions that are performed to transmit the data in an efficient, reliable way over the radio channel: source coding and speech processing (section 4.7), channel coding and burst mapping. Security related functions, such as encryption and authentication are discussed in Chapter 5.

Figure 4.24 gives a schematic overview of the basic elements of the GSM transmission chain. The stream of sampled speech data is fed into a source encoder, which compresses the data by removing unnecessary redundancy (section 4.7). The resulting information bit sequence is passed to the channel encoder (section 4.8). Its purpose is to add, in a controlled manner, some redundancy to the information sequence. This redundancy serves to protect the data against the negative effects of noise and interference encountered in the transmission through the radio channel. On the receiver side, the introduced redundancy allows the channel decoder to detect and correct transmission errors. GSM uses a combination of block and convolutional coding. Moreover, an interleaving scheme is used to deal with burst errors that occur over multipath and fading channels. Next, the encoded and interleaved data are encrypted to guarantee secure and confident data transmission. The encryption technique as well as the methods for subscriber authentication and secrecy of the subscriber identity are explained in section 5.6. The encrypted data are subsequently mapped to bursts (section 4.8.4), which are then multiplexed as explained in previous sections. Finally the stream of bits is differential coded and modulated.

After transmission, the demodulator processes the signal, which was corrupted by the noisy channel. It attempts to recover the actual signal from the received signal. The next steps are demultiplexing and decryption. The channel decoder attempts to reconstruct the

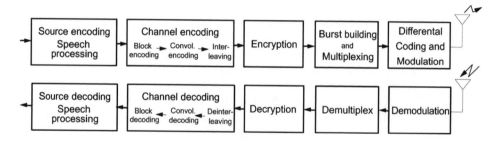

Figure 4.24 Basic elements of GSM transmission chain on the physical layer at the air interface.

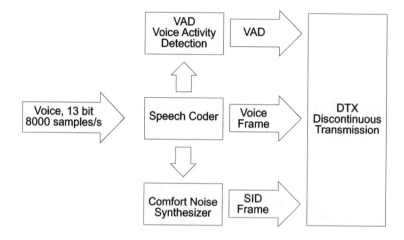

Figure 4.25 Schematic representation of speech functions at the transmitter.

original information sequence and, as a final step, the source decoder tries to reconstruct the original source signal.

4.7 Source coding and speech processing

Source coding reduces redundancy in the speech signal and thus results in signal compression, which means that a significantly lower bit rate is achieved than needed by the original speech signal. The speech coder/decoder is the central part of the GSM speech processing function, both at the transmitter (Figure 4.25) as well as at the receiver (Figure 4.26). The functions of the GSM speech coder and decoder are usually combined in one building block called the codec (COder/DECoder).

The analog speech signal at the transmitter is sampled at a rate of 8000 samples per second and the samples are quantized with a resolution of 13 bits. This corresponds to a bit rate of 104 kbit/s for the speech signal. At the input to the speech codec, a speech frame containing

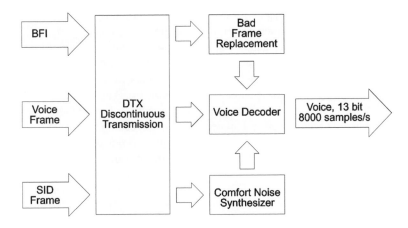

Figure 4.26 Schematic representation of speech functions at the receiver.

160 samples of 13 bits arrives every 20 ms. The speech codec compresses this speech signal into a source-coded speech signal of 260-bit blocks at a bit rate of 13 kbit/s. Thus, the GSM speech coder achieves a compression ratio of 1 to 8. The source coding procedure is briefly explained in the following; detailed discussions of speech coding procedures are given in Steele (1992).

A further ingredient of speech processing at the transmitter is the recognition of speech pauses, called Voice Activity Detection (VAD). The voice activity detector decides, based on a set of parameters delivered by the speech coder, whether the current speech frame (20 ms) contains speech or a speech pause. This decision is used to turn off the transmitter amplifier during speech pauses, under control of the Discontinuous Transmission (DTX) block.

The discontinuous transmission mode takes advantage of the fact, that during a normal telephone conversation, both parties rarely speak at the same time, and thus each directional transmission path has to transport speech data only half of the time. In DTX mode, the transmitter is only activated when the current frame indeed carries speech information. This decision is based on the VAD signal of speech pause recognition. The DTX mode can reduce the power consumption and, hence, prolong the battery life. In addition, the reduction of transmitted energy also reduces the level of interference and thus improves the spectral efficiency of the GSM system. The missing speech frames are replaced at the receiver by a synthetic background noise signal called comfort noise (Figure 4.26). The parameters for the comfort noise synthesizer are transmitted in a special silence descriptor (SID) frame.

This silence descriptor is generated at the transmitter from continuous measurements of the (acoustic) background noise level. It represents a speech frame which is transmitted at the end of a speech burst, i.e. at the beginning of a speech pause. In this way, the receiver recognizes the end of a speech burst and can activate the comfort noise synthesizer with the parameters received in the SID frame. The generation of this artificial background noise in DTX mode prevents the audible background noise transmitted with normal speech bursts from suddenly dropping to a minimal level at a speech pause. This modulation of the background noise would have a very disturbing effect on the human listener and would significantly deteriorate the subjective speech quality. Insertion of comfort noise is a very effective countermeasure to compensate for this so-called noise-contrast effect.

Another loss of speech frames can occur, when bit errors caused by a noisy transmission channel cannot be corrected by the channel coding protection mechanism, and the block is received at the codec as a speech frame in error, which must be discarded. Such bad speech frames are flagged by the channel decoder with the Bad Frame Indication (BFI). In this case, the respective speech frame is discarded and the lost frame is replaced by a speech frame which is predictively calculated from the preceding frame. This technique is called error concealment. Simple insertion of comfort noise is not allowed. If 16 consecutive speech frames are lost, the receiver is muted to acoustically signal the temporary failure of the channel.

The speech compression takes place in the speech coder. The GSM speech coder uses a procedure known as Regular Pulse Excitation Long-Term Prediction (RPE-LTP). This procedure belongs to the family of hybrid speech coders. This hybrid procedure transmits part of the speech signal as the amplitude of a signal envelope, a pure wave form encoding, whereas the remaining part is encoded into a set of parameters. The receiver reconstructs these signal parts through speech synthesis (vocoder technique). Examples of envelope encoding are Pulse Code Modulation (PCM) or Adaptive Delta Pulse Code Modulation (ADPCM). A pure vocoder procedure is Linear Predictive Coding (LPC). The GSM procedure RPE-LTP as well as Code Excited Linear Predictive Coding (CELP) represent mixed (hybrid) approaches (David and Benkner, 1996; Natwig, 1998; Steele, 1992).

A simplified block diagram of the RPE-LTP coder is shown in Figure 4.27. Speech data generated with a sampling rate of 8000 samples per second and 13-bit resolution arrive in blocks of 160 samples at the input of the coder. The speech signal is then decomposed into three components: a set of parameters for the adjustment of the short-term analysis filter (LPC) also called reflection coefficients; an excitation signal for the RPE part with irrelevant portions removed and highly compressed; and finally a set of parameters for the control of the LTP long-term analysis filter. The LPC and LTP analyses supply 36 filter parameters for each sample block, and the RPE coding compresses the sample block to 188 bits of RPE parameters. This results in the generation of a frame of 260 bits every 20 ms, equivalent to a 13 kbit/s GSM speech signal rate.

The speech data preprocessing of the coder (Figure 4.27) removes the DC portion of the signal if present and uses a preemphasis filter to emphasize the higher frequencies of the speech spectrum. The preprocessed speech data are run through a nonrecursive lattice filter (LPC filter; Figure 4.27) to reduce the dynamic range of the signal. Since this filter has a 'memory' of about 1 ms, it is also called short-term prediction filter. The coefficients of this filter, called reflection coefficients, are calculated during LPC analysis and transmitted in a logarithmic representation as part of the speech frame, Log Area Ratios (LARs).

Further processing of the speech data is preceded by a recalculation of the coefficients of the long-term prediction filter (LTP analysis in Figure 4.27). The new prediction is based on the previous and current blocks of speech data. The resulting estimated block is finally subtracted from the block to be processed, and the resulting difference signal is passed on to the RPE coder.

After LPC and LTP filtering, the speech signal has been redundancy reduced, i.e. it already needs a lower bit rate than the sampled signal; however, the original signal can still be reconstructed from the calculated parameters. The irrelevance contained in the speech

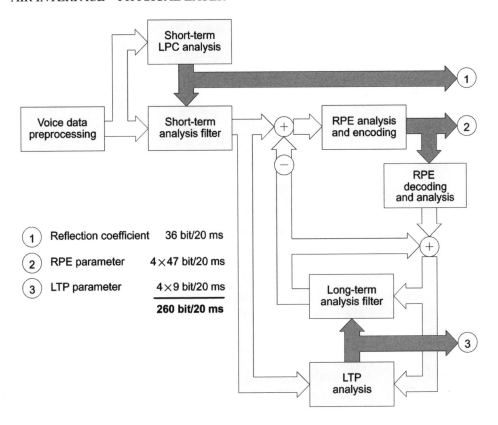

Figure 4.27 Simplified block diagram of the GSM speech coder.

signal is reduced by the RPE coder. This irrelevance represents speech information that is not needed for the understandability of the speech signal, since it is hardly noticeable to human hearing and thus can be removed without loss of quality. On the one hand, this results in a significant compression (factor $160 \times 13/188 \approx 11$); on the other hand, it has the effect that the original signal cannot be reconstructed uniquely. Figure 4.28 summarizes the reconstruction of the speech signal from RPE data, as well as the long-term and short-term synthesis from LTP and LPC filter parameters. In principle, at the receiver site, the functions performed are the inverse of the functions of the encoding process.

The irrelevance reduction only minimally affects the subjectively perceived speech quality, since the main objective of the GSM codec is not just the highest possible compression but also good subjective speech quality. To measure the speech quality in an objective manner, a series of tests were performed on a large number of candidate systems and competing codecs.

The base for comparison used is the Mean Opinion Score (MOS), ranging from MOS = 1, meaning quality is very bad or unacceptable, to MOS = 5, meaning quality is very good or fully acceptable. A series of coding procedures were discussed for the GSM system; they were examined in extensive hearing tests for their respective subjective speech quality (Natwig, 1998). Table 4.10 gives an overview of these test results; it includes as reference

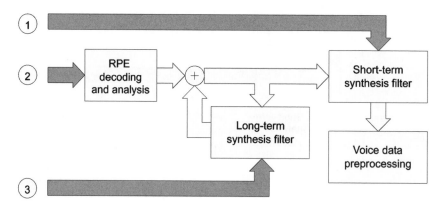

Figure 4.28 Simplified block diagram of the GSM speech decoder.

Table 4.10 MOS results of codec hearing tests (Natwig, 1998).

CODEC	Process	Bit rate (in kbit/s)	MOS
FM	Frequency Modulation	–	1.95
SBC-ADPCM	Subband-CODEC – Adaptive Delta-PCM	15	2.92
SBC-APCM	Subband-CODEC – Adaptive PCM	16	3.14
MPE-LTP	Multi-Pulse Excited LPC-CODEC – Long-Term Prediction	16	3.27
RPE-LPC	Regular-Pulse Excited LPC-CODEC	13	3.54
RPE-LTP	Regular Pulse Excited LPC-CODEC – Long-Term Prediction	13	≈ 4
ADPCM	Adaptive Delta Modulation	32	≥ 4

also ADPCM and frequency-modulated analog transmission. The GSM codec with the RPE-LTP procedure generates a speech quality with an MOS value of about 4 for a wide range of different inputs.

4.8   Channel coding

The heavily varying properties of the mobile radio channel (see Chapter 2) often result in a very high bit error ratio, of the order of 10^{-3} to 10^{-1}. The highly compressed, redundancy-reduced source coding makes speech communication with acceptable quality almost impossible; moreover, it makes reasonable data communication impossible. Suitable error correction procedures are therefore necessary to reduce the bit error probability into an acceptable range of about 10^{-5} to 10^{-6}. Channel coding, in contrast to source coding, adds redundancy to the data stream to enable detection and correction of transmission errors.

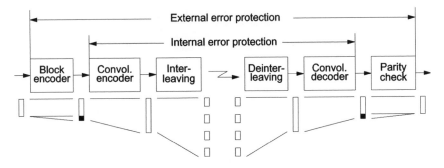

Figure 4.29 Stages of channel coding.

It is the modern high-performance coding and error correction techniques which essentially enable the implementation of a digital mobile communication system.

The GSM system uses a combination of several procedures: in addition to a block code, which generates parity bits for error detection, a convolutional code generates the redundancy needed for error correction. Furthermore, sophisticated interleaving of data over several blocks reduces the damage done by burst errors. The individual steps of channel coding are shown in Figure 4.29:

- calculation of parity bits (block code) and addition of fill bits;

- error protection coding through convolutional coding;

- Interleaving.

Finally, the coded and interleaved blocks are enciphered, distributed across bursts, modulated and transmitted on the respective carrier frequencies.

The sequence of data blocks that arrives at the input of the channel encoder is combined into blocks, partially supplemented by parity bits (depending on the logical channel) and then complemented to a block size suitable for the convolutional encoder. This involves appending zero bits at the end of each data block, which allow a defined resetting procedure of the convolutional encoder (zero termination) and thus a correct decoding decision. Finally, these blocks are run through the convolutional encoder. The ratio of uncoded to coded block length is called the *rate* of the convolutional code. Some of the redundancy bits generated by the convolutional encoder are deleted again for some of the logical channels. This procedure is known as *puncturing*, and the resulting code is a punctured convolutional code (Begin and Haccoun, 1994; Hagenauer *et al.*, 1990; Kallel, 1995). Puncturing increases the rate of the convolutional code, so it reduces the redundancy per block to be transmitted, and lowers the bandwidth requirements, such that the convolution-encoded signal fits into the available channel bit rate. The convolution-encoded bits are passed to the interleaver, which shuffles various bit streams. At the receiving site, the respective inverse functions are performed: deinterleaving, convolutional decoding and parity checking. Depending on the position within the transmission chain (Figure 4.29), one distinguishes between external error protection (block code) and internal protection (convolutional code).

In the following, the GSM channel coding is presented according to these stages. section 4.8.1 explains the block coding, section 4.8.2 deals with convolutional coding and, finally, section 4.8.3 presents the interleaving procedures used in GSM. The error protection measures have different parameters depending on channel and type of transported data. Table 4.11 gives an overview. (Note that the tail bits indicated in the second column are the fill bits needed by the decoding process; they should not be confused with the tail bits of the bursts (section 4.2).)

The basic unit for all coding procedures is the data block. For example, the speech coder delivers a sequence of data blocks to the channel encoder. Depending on the logical channel, the length of the data block is different; after convolutional coding at the latest, data from all channels are transformed into units of 456 bits. Such a block of 456 bits transports a complete speech frame or a protocol message in most of the signaling channels, except for the RACH and SCH channels. The starting points are the blocks delivered to the input of the channel encoder from the protocol processing in higher layers (Figure 4.30).

Speech traffic channels

One block of the full-rate speech codec consists of 260 bits of speech data, i.e. each block contains 260 information bits, which must be encoded. They are graded into two classes (class I, 182 bits; class II, 78 bits) which have different sensitivity against bit errors. Class I includes speech bits that have a greater impact on speech quality and hence must be better protected. Speech bits of class II, however, are less important. They are therefore transmitted without convolutional coding, but are included in the interleaving process. The individual sections of a speech frame are therefore protected to differing degrees against transmission errors (Unequal Error Protection (UEP)). In the case of a half-rate speech codec, data blocks of 112 information bits are input into the channel encoder. Of these, 95 bits belong to class I and 17 bits belong to class II. Again, one data block corresponds to one speech frame.

Data traffic channels

Blocks of traffic channels for data services have a length of N0 bits, the value of N0 being a function of the data service bit rate. We take for example the 9.6 kbit/s data service on a full-rate traffic channel (TCH/F9.6). Here, a bit stream organized into blocks of 60 information bits arrives every 5 ms at the input of the encoder. Four subsequent blocks are combined for the encoding process.

Signaling channels

The data streams of most of the signaling channels are constructed of blocks of 184 bits each; with the exception of the RACH and SCH which supply blocks of length P0 to the channel coder. The block length of 184 bits results from the fixed length of the protocol message frames of 23 octets on the signaling channels. The channel coding process maps pairs of subblocks of 57 bits onto the bursts such that it can fill a normal data burst NB (Figure 4.7).

4.8.1 External error protection: block coding

The block coding stage in GSM has the purpose of generating parity bits for a block of data, which allow the detection of errors in this block. In addition, these blocks are supplemented by fill bits (tail bits) to a block length suitable for further processing. Since block coding is the

Table 4.11 Error protection coding and interleaving of logical channels.

Channel type	Abbreviation	Block distance (ms)	Bits per block			Convolution code rate	Encoded bits per block	Interleaver depth
			Data	Parity	Tail			
TCH, full rate, speech	TCH/FS	20	260				456	8
Class I			182	3	4	1/2	378	
Class II			78	0	0	–	78	
TCH, half rate, speech	TCH/HS	20	112				228	4
Class I			95	3	6	104/211	211	
Class II			17	0	0	–	17	
TCH, full rate, 14.4 kbit/s	TCH/F14.4	20	290	0	4	294/456	456	19
TCH, full rate, 9.6 kbit/s	TCH/F9.6	5	4×60	0	4	244/456	456	19
TCH, full rate, 4.8 kbit/s	TCH/F4.8	10	60	0	16	1/3	228	19
TCH, half rate, 4.8 kbit/s	TCH/H4.8	10	4×60	0	4	244/456	456	19
TCH, full rate, 2.4 kbit/s	TCH/F2.4	10	2×36	0	4	1/6	456	8
TCH, half rate, 2.4 kbit/s	TCH/H2.4	10	2×36	0	4	1/3	228	19
FACCH, full rate	FACCH/F	20	184	40	4	1/2	456	8
FACCH, half rate	FACCH/H	40	184	40	4	1/2	456	6
SDCCH, SACCH			184	40	4	1/2	456	4
BCCH, NCH, AGCH, PCH		235	184	40	4	1/2	456	4
RACH		235	8	6	4	1/2	36	1
SCH			25	10	4	1/2	78	1
CBCH		235	184	40	4	1/2	456	4

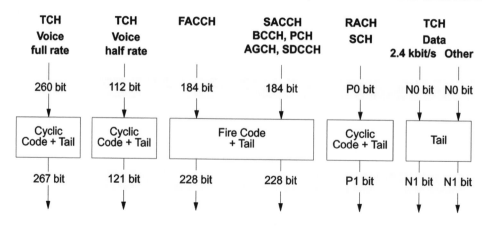

Figure 4.30 Overview of block coding for logical channels (also see Table 4.11).

first or external stage of channel coding, the block code is also known as external protection. Figure 4.30 gives a brief overview showing which codes are used for which channels. In principle, only two kinds of codes are used: a Cyclic Redundancy Check (CRC) and a fire code.

Block coding for speech traffic channel

As mentioned above, speech data occurs on the TCH in speech frames (blocks) of 260 bits for TCH/F and 112 bits for TCH/H, respectively. The bits belonging to class I are error-protected, whereas the bits of class II and are not protected. A 3-bit CRC code is calculated for the first 50 bits of class I (in the case of TCH/F). The generator polynomial for this CRC is

$$G_{CRC}(x) = x^3 + x + 1.$$

In the case of a TCH/H speech channel, the most significant 22 bits of Class I are protected by 3 parity bits, using the same generator polynomial.

We now explain the block coding process in more detail with focus on the TCH/F speech codec. Since cyclic codes are easily generated with a feedback shift register, they are often defined directly with this register representation. Figure 4.31 shows such a shift register with storage locations (delay elements) and modulo-2 adders. For initialization, the register is primed with the first three bits of the data block. The other data are shifted bitwise into the feedback shift register; after the last data bit has been shifted out of the register, the register contains the check sum bits, which are then appended to the block.

The operation of this shift register can be easily explained if the bit sequences are also represented as polynomials as with the generating function. The first 50 bits of a speech frame $D_0, D_1, \ldots, D_{49}$ are denoted as

$$D(x) = D_{49}x^{49} + D_{48}x^{48} + \cdots + D_1x + D_0.$$

If this data sequence is shifted through the register of Figure 4.31, after the register was primed with D_{47}, D_{48}, D_{49} followed by 50 shift operations, then the check sum bits

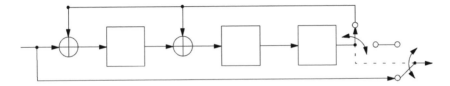

Figure 4.31 Feedback shift register for CRC.

$R(x)$ correspond to the remainder, which is left by dividing the data sequence $x^3 D(x)$ (supplemented by three zero bits) by the generator polynomial:

$$R(x) = \text{Remainder}\left[\frac{x^3 D(x)}{G_{CRC}(x)}\right].$$

In the case of error-free transmission, the codeword $C'(x) = x^3 D(x) + R(x)$ is therefore divisible by $G_{CRC} C(x)$ without remainder. However, since the check sum bits $R(x)$ are transmitted in inverted form, the division yields a remainder:

$$S(x) = \text{Remainder}\left[\frac{C(x)}{G_{CRC}(x)}\right] = \text{Remainder}\left[\frac{x^3 D(x) + \bar{R}(x)}{G_{CRC}(x)}\right] = x^2 + x + 1.$$

This is equivalent to shifting the whole codeword $C(x)$ through an identical shift register on the decoder side, after priming it with C_{50}, C_{51}, C_{52}. After shifting in the last check sum bit (50 shift operations), this register should contain a 1. If this is not the case, the block contains erroneous bits. Inversion of the parity bits avoids the generation of null code-words, i.e. bursts which contain only zeros cannot occur on the traffic channel.

The speech data $d(k)$ ($k = 1, \ldots, 182$) of class I of a block are combined with the parity bits $p(k)$ ($k = 1, 2, 3$) and fill bits to form a new block $u(k)$ ($k = 1, \ldots, 189$)

$$u(k) = \begin{cases} d(2k) & k = 1, \ldots, 90, \\ d(2 \times (184 - k) + 1) & k = 94, \ldots, 184, \\ p((k - 91) + 1) & k = 91, 92, 93, \\ 0 & k = 185, \ldots, 189. \end{cases}$$

The bits in even or odd positions are shifted to the upper or lower half of the block, respectively, and separated by the three check sum bits; in addition, the order of the odd bits is reversed. Finally, the block is filled to 189 bits. Combination with the speech bits of class II yields a block of 267 bits, which serves as input to the convolutional coder.

This enormous effort is taken because of the high compression rate and sensitivity against bit errors of the speech data. A speech frame in which the bits of class I have been recognized as erroneous can therefore be reported as erroneous to the speech codec using the BFI; see section 4.7. In order to maintain a constantly good speech quality, speech frames recognized as faulty are discarded, and the last correctly received frame is repeated, or an extrapolation of received speech data is performed.

Table 4.12 Block formation for data traffic channels.

Data channel	N0		Tail bits		N1
TCH/F14.4	290	+	4	=	294
TCH/F9.6	4 × 60	+	4	=	244
TCH/F4.8	(2 ×)60	+	(2 ×)16	=	(2 ×)76
TCH/H4.8	4 × 60	+	4	=	244
TCH/F2.4	2 × 36	+	4	=	76
TCH/H2.4	(2 ×)2 × 36	+	(2 ×)4	=	(2 ×)76

Block coding for data traffic channels

Block coding of traffic channels is somewhat simpler for data services. In this case, no parity bits are determined. Blocks of length N0 arriving at the input of the encoder are supplemented by fill bits to a size of N1 suitable for further coding. Table 4.12 gives an overview of the different block lengths, which depend on the data rate and channel type, i.e. whether the channel is a full-rate (TCH/Fxx) or half-rate (TCH/Hxx) channel.

The 9.6 kbit/s data service is only offered on a full-rate traffic channel. The data comes in blocks of 60 bits to the channel encoder (every 5 ms). Four blocks each are combined and supplemented by four appended tail bits (zero bits). In the case of a nontransparent data service, these four blocks make up exactly one protocol frame of the Radio Link Protocol (RLP; 240 bits). The procedures for other data services are similar. As shown in Table 4.11, for the 4.8 and 2.4 kbit/s services, blocks of 60 or 36 bit length arrive every 10 ms. Subsequent blocks are combined and are then supplemented with tail bits (zero bits) to form blocks of 76 or 244 bits, respectively. The bit stream for the 14.4 kbit/s data service (TCH/F14.4) is offered to the encoder in blocks of 290 information bits every 20 ms. Here, four tail bits are added, resulting in 294 bits (see Table 4.12).

Block coding for signaling channels

The majority of the signaling channels (SACCH, FACCH, SDCCH, BCCH, PCH and AGCH) use an extremely powerful block code for error detection. This is a so-called fire code, i.e. a shortened binary cyclic code which appends 40 redundancy bits to the 184-bit data block. Its pure error-detection capability is sufficient to let undetected errors go through only with a probability of 2^{-40}. (A fire code can also be used for error correction, but here it is used only for error detection.) Error detection with the fire code in the SACCH channel is used to verify connectivity (Figure 4.23), and is used, if indicated, for decisions regarding breaking a connection. The fire code can be defined in the same way as the CRC by way of a generator polynomial:

$$TG_F(x) = (x^{23} + 1)(x^{17} + x^3 + 1).$$

The check sum bits $R_F(s)$ of this code are calculated in such a way that a 40-bit remainder $S_F(x)$ is left after dividing the codeword $C_F(x)$ by the generator polynomial $G_F(x)$. In the

Table 4.13 Block lengths for the RACH and SCH channels.

Data channel	P0	Parity bits	Tail bits	P1
RACH	8	+6	+4	= 18
SCH	25	+10	+4	= 39

case of no errors, the remainder contains only '1' bits:

$$S_F(x) = \text{Remainder}\left[\frac{C_F(x)}{G_F(x)}\right] = \text{Remainder}\left[\frac{x^{40} D_F(x) + R_F(x)}{G_F(x)}\right]$$

$$= x^{39} + x^{38} + \cdots + x^2 + x + 1.$$

The codeword generated with the redundancy bits of the fire code is supplemented with '0' bits to a total length of 228 bits, which are then delivered to the convolutional coder.

Another approach has been used for error detection in the RACH channel. The very short random access burst in the RACH allows only a data block length of P0 = 8 bits, which is supplemented in a cyclic code by six redundancy bits. The corresponding generator polynomial is

$$G_{RACH}(x) = x^6 + x^5 + x^3 + x^2 + x + 1.$$

In the AB, the MS also has to indicate a target base station. The BSIC of the respective base station is used for this purpose. The six bits of the BSIC are added to the six redundancy bits modulo-2, and the resulting sequence is inserted as the redundancy of the data block. The total codeword to be convolution-coded for the RACH thus has a length of 18 bits; i.e. four fill bits ('0') are also added in the RACH to this block. In exactly the same way, block coding is performed for the handover access burst, which is in principle also a random access burst.

The SCH channel, as an important synchronization channel, uses somewhat more elaborate error protection than the RACH channel. The SCH data blocks have a length of 25 bits and receive, in addition to the fill bits, another 10 bits of redundancy for error detection through a cyclic code with somewhat better error detection capability than on the RACH:

$$G_{SCH}(x) = x^{10} + x^8 + x^6 + x^5 + x^4 + x^2 + 1.$$

Thus the length of the codewords delivered to the channel coder in the SCH channel is 39 bits. Table 4.13 summarizes the block parameters of the RACH and SCH channels. Table 4.14 presents an overview of the cyclic codes used in GSM.

4.8.2 Internal error protection: convolutional coding

After block coding has supplemented the data with redundancy bits for error detection (parity bits), added fill bits and thus generated sorted blocks, the next stage is the calculation of additional redundancy for error correction to correct the transmission errors caused by the radio channel. The internal error correction of GSM is based exclusively on convolutional codes.

Convolutional codes (Johannesson and Zigangirov, 1999) can also be defined using shift registers and generator polynomials. Figure 4.32 illustrates a possible convolutional encoder

Table 4.14 Cyclic codes used for block coding in GSM.

Channel	Polynomial
TCH/FS	$x^3 + x + 1$
DCCH and CCCH (part)	$(x^{23} + 1)(x^{17} + x^3 + 1)$
RACH	$x^6 + x^5 + x^3 + x^2 + x + 1$
SCH	$x^{10} + x^8 + x^6 + x^5 + x^4 + x^2 + x + 1$

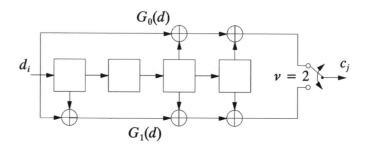

Figure 4.32 Principle of a convolutional encoder.

realization. It basically consists of a shift register with modulo-2 adders and K storage locations (here $K = 4$). One data/information symbol d_i is read into the shift register per tact interval. A symbol consists of k (here $k = 1$) data/information bits, each of which is moved into the shift register. A data symbol could also consist of more than one bit ($k > 1$), but this is not implemented in GSM. The symbol read is combined with up to K of its predecessor symbols $d_{i-1}, \ldots, d_{i-K}$ in several modulo-2 additions. The results of these operations are given to the interleaver as coded user payload symbols c_j. The value K determines the number of predecessor symbols to be combined with a data symbol and is therefore also called the memory of the convolutional encoder. The number v of combinatorial rules (here $v = 2$) determines the number of coded bits in a code symbol c_j generated for each input symbol d_i. In Figure 4.32, the combinatorial results are scanned from top to bottom to generate the code symbol c_j. The combinatorial rules are defined by the generator polynomial $G_i(d)$. It is important to note that a specific convolutional code can be generated by various encoders. Thus, it must be carefully distinguished between code properties and encoder properties.

As mentioned in section 4.8.1, block coding appends at least four zero bits to each block. These bits not only serve as fill bits at the end of a block, but they are also important for the channel coding procedure. Shifted at the end of each block into the encoder, these bits serve to reset the encoder into the defined starting position (zero termination of the encoder), such that in principle adjacent data blocks can be coded independently of each other.

The rate r of a convolutional code indicates how many data (information) bits are processed for each coded bit. Consequently, $1/r$ is the number of coded bits per information bit. This rate is the essential measure of the redundancy produced by the code and, hence, its

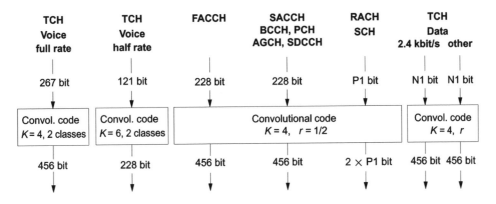

Figure 4.33 Overview of convolutional coding of logical channels (also see Table 4.11).

error correction capability:

$$r = \frac{k}{v}, \quad \text{here:} \quad r = \frac{1}{v} = \frac{1}{2}.$$

The code rate is therefore determined by the number of bits k per input data symbol and the number of combinatorial rules v which are used for the calculation of a code symbol. In combination with the memory K, the code rate r determines the error-correction capability of the code. In a simplified way: with decreasing r and increasing K, the number of corrigible errors per codeword increase, and, thus, the error-correction capabilities of the code are improved. The encoding procedure is expressed in the combinatorial operations (modulo-2 additions). These coding rules can be described with polynomials. In the case of the convolutional encoder of Figure 4.32, the two generator polynomials are

$$G_0(d) = d^4 + d^3 + 1,$$
$$G_1(d) = d^4 + d^3 + d + 1.$$

These give a compact representation of the encoding procedure. The maximal exponent of a generator polynomial is known as its constraint length. The maximal of all constraint lengths (i.e. the maximal exponent of all polynomials) defines the memory K of the convolutional encoder. The number of polynomials determines the rate r. The exponents represent how an input symbol d_i is processed in the encoder. For example, in the upper path of the encoder (represented by $G_0(d)$), an input symbol d_i is immediately forwarded to the output (exponent '0'), and it is processed again in the third and fourth tact interval.

GSM defines different convolutional codes for the different logical channels (see Figure 4.33). Table 4.15 lists the seven generator polynomials ($G_0, \ldots, G_6$) used in different combinations. The convolutional encoder used for half-rate speech channels (TCH/HS) has memory 6. All other encoders have memory 4, but they differ in the code rate and the polynomials used.

Table 4.16 gives an overview of the uses and combinations of generator polynomials. Most logical channels use a convolutional code of rate 1/2 based on polynomials G0 and G1.

Table 4.15 Generator polynomials for convolutional codes.

Type	Polynomial
G0	$1 + d^3 + d^4$
G1	$1 + d + d^3 + d^4$
G2	$1 + d^2 + d^4$
G3	$1 + d + d^2 + d^3 + d^4$
G4	$1 + d^2 + d^3 + d^5 + d^6$
G5	$1 + d + d^4 + d^6$
G6	$1 + d + d^2 + d^3 + d^4 + d^6$

Table 4.16 Usage of generator polynomials.

Channel type	Generator polynomial						
	G0	G1	G2	G3	G4	G5	G6
TCH, full rate, speech							
Class I	■	■					
Class II							
TCH, half rate, speech							
Class I					■	■	■
Class II							
TCH, full rate, 14.4 kbit/s	■	■					
TCH, full rate, 9.6 kbit/s	■	■					
TCH, full rate, 4.8 kbit/s	■	■	■				
TCH, half rate, 4.8 kbit/s	■	■					
TCH, full rate, 2.4 kbit/s					■		
TCH, half rate, 2.4 kbit/s					■		
FACCHs	■	■					
SDCCHs, SACCHs	■	■					
BCCH, AGCH, PCH	■	■					
RACH	■	■					
SCH	■	■					

Speech traffic channels

Convolutional coding of class I speech bits on the full-rate speech channel generates $1/r \times (182 + 3 + 4)$ bits $= 378$ bits. The 78 bits of class II are not encoded at all, which results in a total of 456 bits. This is the uniform block size needed for mapping these data blocks onto the bursts with 114-bit payload. In a similar way, for most of the remaining channels, two coded blocks of 228 bits are combined.

In the case of a half-rate speech traffic channel (TCH/HS), the class I bits are encoded using a punctured version of a rate-1/3 convolutional encoder defined by G4, G5 and G6. Including puncturing, the net rate of the encoder is $r' = 104/211 \approx 1/2$. The 95 information bits, 3 parity bits and 6 tail bits are thus mapped to 211 bits. The 17 class II bits are not convolutional encoded, which results in a total number of 228 bits.

Data traffic channels

The 4.8 kbit/s data service on a full-rate channel (TCH/F4.8) and the 2.4 kbit/s data service on a half-rate channel (TCH/H2.4) use a code of rate 1/3 based on polynomials G1, G2 and G3. The 2.4 kbit/s data service on a full-rate channel (TCH/F2.4) uses these three polynomials twice in a row to generate a convolutional code of rate 1/6.

The data on a TCH/F9.6 and TCH/H4.8 is encoded using the rate-1/2 code defined by the polynomials G0 and G1. At the input of the convolution encoder, blocks of 244 bits arrive which the encoder maps to blocks of 488 bits. These blocks are reduced to 456 bits by removing (puncturing) every 15th bit beginning with the 11th bit, i.e. a total of 32 bits are punctured. On the one hand, puncturing cuts the block size to a length suitable for further processing; on the other hand, puncturing removes redundancy. The resulting net code rate of $r' = 244/456$ is therefore somewhat higher than the rate of 1/2 for the convolutional encoder. Thus, the code has slightly lower error-correction capability. Puncturing cuts down the convolutional coded blocks of the TCH/F9.6 and TCH/H4.8 channels to the standard format of 456 bits. Thus, blocks of these channels can also be processed in a standardized way (interleaving, etc.), and the amount of redundancy contained in a block is also matched to the bit rate available for transmission.

For the encoding of the TCH/F14.4, again a punctured version of the (G0, G1) convolutional encoder is employed. The 294 bits at the input of the encoder are mapped to 588 bits, followed by a puncturing of 132 bits.

Convolutional decoding

In most cases, the decoding of a convolutional code employs the Viterbi algorithm. It uses a suitable metric to determine the data sequence that most likely equals the transmitted data (maximum likelihood decoding) (Bossert, 1999). Using the knowledge of the generator polynomials, the decoder can determine the original data sequence.

4.8.3 Interleaving

The decoding result of the convolutional code is highly dependent on the frequency and grouping of bit errors that occur during transmission. In particular, burst errors during long and deep fading periods, i.e. a series of erroneous sequential bits, have a negative impact on error correction. In such cases, the channel is not a binary channel without memory, rather the single-bit errors have statistical dependence, which diminishes the result of the error-correction procedure of the convolutional code. To achieve good error correction results, the channel should have no memory, i.e. the bit errors should be statistically independent. Therefore, burst errors occurring frequently on the radio channel should be distributed uniformly across the transmitted codewords. This can be accomplished through the interleaving technique described in the following.

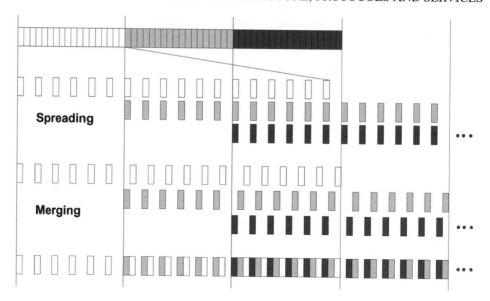

Figure 4.34 Interleaving: spreading and merging.

The interleaving approach is to distribute codewords from the convolutional encoder by spreading in time and merging them across several bursts for transmission. This principle is shown in Figure 4.34. By time spreading, each of the codewords is distributed across a threefold length. Merging the bit sequences generated in this way has the effect that the individual bits from each of the three codewords are sorted into alternate bursts; this way each codeword is transmitted as distributed over a total of three bursts, and two bits of a data block are never transmitted adjacent to each other.

This kind of interleaving is also known as diagonal interleaving. The number of bursts over which a codeword is spread is called the interleaving depth; a spreading factor can be defined analogously. A burst error is therefore distributed uniformly over several subsequently transmitted codewords because of the distribution of the data over several bursts. This generates bit error sequences which are less dependently distributed in the data stream, hence it improves the success of the error-correction process.

Figure 4.35 shows an example. During the third burst of transmission, severe fading of the signal leads to a massive burst error. This burst is now heavily affected by a total of six single-bit errors. In the process of deinterleaving (inversion of merging, despreading) these bit errors are distributed across three data blocks, corresponding to the bit positions which were sorted into the respective bursts during interleaving. The number of errors per data block is now only two, which can be much more easily corrected.

Another kind of interleaving is block interleaving. In this principle, codewords are written line by line into a matrix (Figure 4.36), which is subsequently read out column by column. The number of lines of the interleaving matrix determines the interleaving depth. As long as the length of a burst error is shorter than the interleaving depth, the burst error generates only single-bit errors per codeword if block interleaving is used (Bossert, 1999; Steele, 1992).

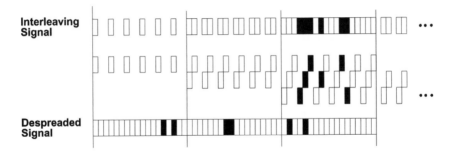

Figure 4.35 Distributing bit errors through deinterleaving.

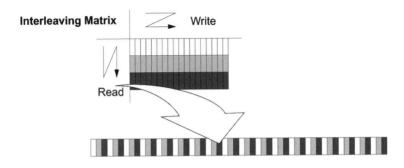

Figure 4.36 Principle of block interleaving.

However, the great advantage of interleaving, to alleviate the effect of burst errors for optimal error correction with a convolutional code, is traded for a significant disadvantage for speech and data communication. As evident from Figures 4.34 and 4.36, the bits of a codeword are spread across several bursts (three in the example here). For a complete reconstruction of a codeword, one has to wait for the complete transmission of three bursts. This forces a transmission delay, which is a function of the interleaving depth.

In GSM, both methods of interleaving are used (Figure 4.37), blockwise as well as bitwise. With a maximal interleaving depth of 19, this can lead to delays of up to 360 ms (Table 4.17).

Full-rate speech channel, TCH/F2.4 and FACCH

The speech channel TCH/FS in GSM uses block-diagonal interleaving. The 456 bits of a codeword are distributed across eight interleaving blocks, where one interleaving block has 114 bit positions. The exact interleaving rule for mapping the coded bits $c(n, k = 0, \ldots, 455)$ of the nth codeword, onto bit position $i(b, j = 0, \ldots, 114)$ of the bth interleaving block, is

$$i(b, j) = c(n, k),$$

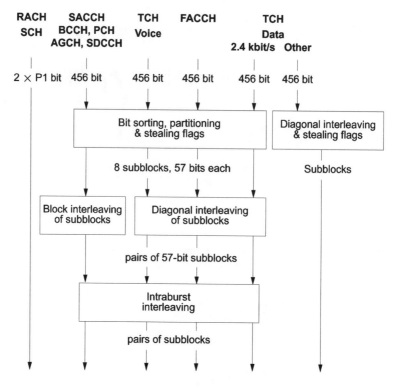

Figure 4.37 Overview of the interleaving of (full-rate) logical channels.

with

$$
\begin{cases}
n = 0,\ 1,\ 2,\ \dots,\ N,\ N + 1,\ \dots, \\
k = 0,\ 1,\ 2,\ \dots,\ 455, \\
b = b_0 + 4n + (k \bmod 8), \\
j = 2((49k) \bmod 57)) + ((k \bmod 8)\ \mathrm{div}\ 4).
\end{cases}
$$

The bits of the nth codeword (data block n in Figure 4.38) are distributed across eight interleaving blocks, beginning with block $B = b_0 + 4n$. To do so, the coded bits are mapped to the even bits of the first four interleaving blocks $(B + 0, \dots, B + 3)$ and to the odd bits of the other four interleaving blocks $(B + 4, \dots, B + 7)$. The even bits of the last four interleaving blocks $(B + 4, \dots, B + 7)$ are occupied by data from codeword $n + 1$. Each interleaving block thus contains 57 bits of the current codeword n and 57 bits of the following codeword $n + 1$ or the preceding codeword $n - 1$, respectively. In this way, a new codeword is started after each fourth merged interleaving block.

The individual bits of codeword n are alternatively distributed across the interleaving blocks, e.g., every eighth bit is in the same interleaving block according to the term $(k \bmod 8)$, whereas bit position j within an interleaving block $b = B + 0, B + 1, \dots, B + 7$ is determined by two terms: the term $(k \bmod 8)\ \mathrm{div}\ 4$ is used to determine the even/odd bit positions; and the term $2\,((49k) \bmod 57)$ determines the offset within the interleaving

Table 4.17 Transmission delay caused by interleaving.

Channel type	Interleaving depth	Transmission delay (ms)
TCH, full rate, voice	8	38
TCH, half rate, voice	4	
TCH, full rate, 14.4 kbit/s	19	93
TCH, full rate, 9.6 kbit/s	19	93
TCH, full rate, 4.8 kbit/s	19	93
TCH, half rate, 4.8 kbit/s	19	185
TCH, full rate, 2.4 kbit/s	8	38
TCH, half rate, 2.4 kbit/s	19	185
FACCH, full rate	8	38
FACCH, half rate	8	74
SDCCH	4	14
SACCH/TCH	4	360
SACCH/SDCCH	4	14
BCCH, AGCH, PCH	4	14

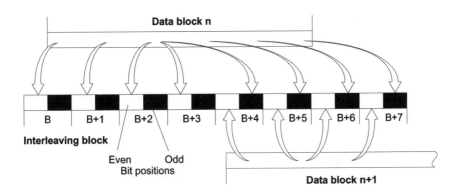

Figure 4.38 Interleaving TCH/FS: block mapping.

block. The first interleaving block B derived from codeword n thus contains bit numbers $0, 8, 16, \ldots, 448, 456$ of this codeword.

The placement of these bits for the first block B in the interleaving block is illustrated in Figure 4.39. This placement is chosen in such a way that no two directly adjacent bits of the interleaving block belong to the same codeword. In addition, the mapped bits are combined into groups of eight bits each, which are distributed as uniformly as possible across the entire interleaving block. This achieves additional spreading of error bursts within a data block. Therefore, the interleaving for the TCH/FS is block-diagonal interleaving with additional merging of data bits within the interleaving block. This is also called intraburst interleaving

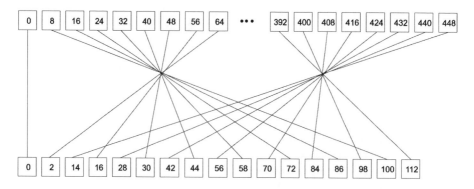

Figure 4.39 Mapping of codeword n onto interleaving block B for a TCH/FS.

(Figure 4.37). The data channel TCH/F2.4 and the FACCH in GSM use the same interleaving methods as the TCH/FS.

Other data traffic channels

For the other data services in the traffic channel (TCH/F14.4, TCH/F9.6, TCH/F4.8, TCH/H4.8 and TCH/H2.4) the interleaving is somewhat simpler. A pure bitwise diagonal interleaving with an interleaving depth of 19 is used. In this case, the interleaving rule is

$$i(b, j) = c(n, k),$$

with

$$\begin{cases} n = 0, 1, 2, \ldots, N, N + 1, \ldots, \\ k = 0, 1, 2, \ldots, 455, \\ b = b_0 + 4n + (k \bmod 19) + k, \\ j = k \bmod 19 + 19((k \bmod 6) \operatorname{div} 114. \end{cases}$$

The bits of a data block (n, k) are distributed in groups of 114 bits across 19 interleaving blocks, whereby groups of six bits are distributed uniformly over one interleaving block.

With this diagonal interleaving, each interleaving block also starts a new 114-bit block of data. A closer look at this interleaving rule reveals that the input to the interleaver consists of blocks of 456 coded data bits as codewords. The whole codeword is therefore really spread across 22 interleaving blocks; the nominal interleaving depth of 19 results historically from 114-bit block interleaving.

Half-rate speech channel

The interleaving rule for the half-rate speech channel (TCH/HS) is given by

$$i(b, j) = c(n, k),$$

with

$$\begin{cases} n = 0, 1, 2, \ldots, N, N + 1, \ldots, \\ k = 0, 1, 2, \ldots, 227, \\ b = b_0 + 2n + (k \bmod 4), \end{cases}$$

and j according to a table in the GSM standard. The 228 bits of a codeword n are distributed over 4 blocks. Beginning with interleaving block $B = b_0 + 2n$, it occupies the even-numbered bits of the first two interleaving blocks ($B + 0$, $B + 1$) and the odd-numbered bits of the other two blocks ($B + 2$, $B + 3$). Consequently, the following codeword $n + 1$ uses the even-numbered bits of the blocks $B + 2 (= b_0 + 2(n + 1) + 0)$ and $B + 3(= b_0 + 2(n + 1) + 1)$ as well as the odd-numbered bits of the interleaving blocks $b_0 + 2(n + 1) + 2$ and $b_0 + 2(n + 1) + 3$. As with the TCH/FS, one interleaving block contains 57 bits from codeword n and 57 bits from codeword $n + 1$ or $n - 1$. In summary, a new codeword starts every second interleaving block.

Signaling channels

Most signaling channels use an interleaving depth of 4, such as SACCH, BCCH, PCH, AGCH and SDCCH. The interleaving scheme is almost identical to that used for the TCH/FS; however, the codewords $c(n, k)$ are spread across four rather than eight interleaving blocks:

$$i(b, j) = c(n, k),$$

with

$$\begin{cases} n = 0, 1, 2, \ldots, N, N + 1, \ldots, \\ k = 0, 1, 2, \ldots, 455, \\ b = b_0 + 4n + (k \bmod 4), \\ j = 2((49k) \bmod 57)) + ((k \bmod 8) \text{ div } 4). \end{cases}$$

With this kind of interleaving, there are also eight blocks generated, just as in the case of the TCH/FS; however, at the same time a block of 57 even bits is combined with a block of 57 odd bits to form a complete interleaving block. This has the consequence that consecutive coded signaling messages are not block-diagonally interleaved, but that each four consecutive interleaving blocks are fully occupied with the data of just one, and only one, codeword. Also, a new codeword starts after every four interleaving blocks. Therefore, this interleaving of GSM signaling messages is in essence also a block interleaving procedure. This is especially important for signaling channels to ensure the transmission of individual protocol messages independent of preceding or succeeding messages. This also enables some kind of asynchronous communication of signaling information. The signaling data of the RACH and SCH must each be transmitted in single data bursts; no interleaving occurs.

4.8.4 Mapping onto the burst plane

After block encoding, convolutional encoding and interleaving, the data are available in the form of 114-bit interleaving blocks. This corresponds exactly to the amount of data which can be carried by a NB (Figure 4.7). Each interleaving block is mapped directly onto one burst (Figure 4.40). After setting the stealing flags, the bursts can be composed and passed to the

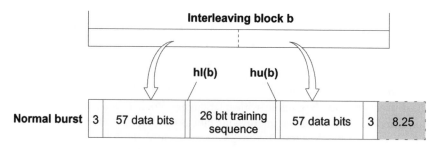

Figure 4.40 Mapping onto a burst.

modulator. The stealing flags indicate whether high-priority signaling messages are present (FACCH messages), which must be transmitted as fast as possible, instead of the originally planned data of the traffic channel.

An essential component of GSM channel coding is the correct treatment of FACCH signaling messages which are multiplexed in a preemptive way into the traffic channel. At the burst level, each FACCH codeword displaces a codeword of the current TCH/FS traffic channel, i.e. the codewords must be tied into the interleaving structure instead of regular data blocks of the traffic channel. The interleaving rule for the FACCH is the same as for the TCH/FS: from an FACCH codeword the even positions are occupied in one set of four interleaving blocks, and the odd bit positions are occupied in another set of interleaving blocks; in addition, the bit positions within the interleaving blocks are shuffled (intra-burst interleaving).

When an FACCH message needs to be transmitted (e.g. a handover command), the current data block n is replaced by the convolutional coded FACCH message, and it is interleaved in a block-diagonal way with the data blocks $(n-1)$ and $(n+1)$ of the traffic channel (Figure 4.41). In the eight blocks involved in this procedure, $B, B+1, \ldots, B+7$, the respective stealing flags hl(b) and hu(b) have to be set (Figure 4.40). If neither flag is set, the burst contains data of the traffic channel. If the even bits of the burst are occupied by FACCH data, hu(b) is set; in the case of the odd bits being used for FACCH, hl(b) is set (Figure 4.41).

If the current burst is not available for traffic channel data, the data block n has to be discarded. Bits 'stolen' in this way have various effects.

- A complete speech frame of the TCH/FS is lost (20 ms speech).

- In the case of TCH/F9.6 and TCH/H4.8 channels, 3 bits are stolen from each of the 8 interleaving blocks, which belong to the same data block, such that a maximum of 24 coded data bits are interfered with.

- In the case of TCH/F4.8 and TCH/H2.4 channels, 6 bits are stolen from each of the 8 interleaving blocks, which belong to the same data block, such that a maximum of 48 coded data bits are interfered with.

- In the case of TCH/F2.4 channels, the same interleaving rules as in the TCH/FS are used, such that a complete data block is displaced by the FACCH.

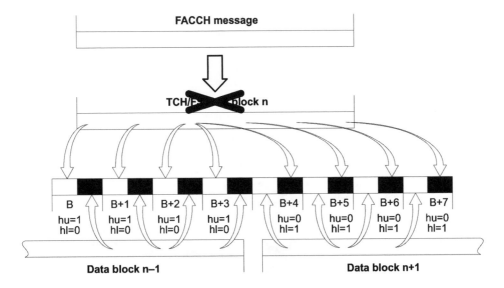

Figure 4.41 Insertion of an FACCH message into the TCH/FS data stream.

In summary, the FACCH signaling needed for fast reactions causes data losses or bit errors in the accompanying traffic channel, and they have to be totally or partially corrected by the convolutional code.

4.8.5   Improved codecs for speech services: half-rate codec, enhanced full-rate codec and adaptive multi-rate codec

One of the most important services in GSM is (of course) voice service. Thus, it is obvious that the voice service has to be further improved. The priority is the development of new speech codecs with two competing objectives:

- better utilization of the frequency bands assigned to GSM; and

- improvement of speech quality to a level that is similar to that offered by ISDN networks, which is the primary request of professional users.

Half-rate codec

The reason for improved bandwidth utilization is to increase the network capacity and the spectral efficiency (i.e. traffic carried per cell area and frequency band). Early plans were already in place to introduce a *half-rate speech codec*. Under good channel conditions, this codec achieves, in spite of the half bit rate, almost the same speech quality as the full-rate codec used so far. However, quality loss occurs in particular for mobile-to-mobile communication, since in this case (due to the ISDN architecture) one has to go twice through the GSM speech coding/decoding process. These multiple, or tandem, conversions degrade speech quality. The end-to-end transmission of GSM-coded speech is intended to avoid

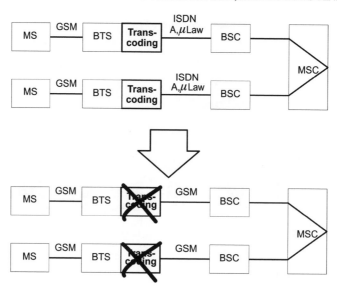

Figure 4.42   Through-transport of GSM-coded speech in Phase 2+ for mobile-to-mobile connections (TFO).

multiple unnecessary transcoding and the resulting quality loss (Figure 4.42) (Mouly and Pautet, 1995). This technique has been passed under the name Tandem Free Operation (TFO) in GSM Release 98.

Enhanced full-rate codec

A very important concern is the improvement of speech quality. Speech quality that is close to what is found in fixed networks is particularly important for business applications and in cases where GSM systems are intended to replace fixed networks, e.g., for fast installation of telecommunication networks in areas with insufficient or missing telephone infrastructure.

Work on the Enhanced Full-Rate (EFR) codec was therefore considered as high priority. This EFR is a full-rate codec (net bit rate 12.2 kbit/s). Nevertheless, it achieves speech quality that is clearly superior to the previously used full-rate codec. It has been initially standardized and used in North American DCS1900 networks (Mouly and Pautet, 1995) and has been implemented in GSM with very good success. Instead of using the RPE-LTP coding scheme (section 4.7), a so-called Algebraic Code Excitation – Linear Prediction (ACELP) is employed.

The EFR speech coder delivers data blocks of 244 information bits to the channel encoder (compare with Table 4.11). In addition to grading the bits into important class I bits and less important class II bits, EFR further divides into class Ia bits and class Ib bits. A special preliminary channel coding is employed for the most significant bits: eight parity bits (generated by a CRC) and eight repetition bits are added to provide additional error detection. The resulting 260 bits are processed by the block encoder as described in section 4.8.1.

For convolutional coding of class I bits the convolutional encoder defined by the generator polynomials G0 and G1 is employed.

Adaptive multi-rate codec

The speech codecs mentioned before (full rate, half rate and EFR) all use a fixed source/information bit rate, which has been optimized for typical radio channel conditions. The problem with this approach is its inflexibility: whenever the channel conditions are much worse than usual, very poor speech quality will result, since the channel capacity assigned to the MS is too small for error free transmission. On the other hand, radio resources will be wasted for unneeded error protection if the radio conditions are better than usual.

To overcome these problems, a much more flexible codec has been developed and standardized: the Adaptive Multi-rate (AMR) codec. It can improve speech quality by adaptively switching between different speech coding schemes (with different levels of error protection) according to the current channel quality. To be more precise, AMR has two principles of adaptability (Bruhn *et al.*, 2000): channel mode adaptation and codec mode adaptation.

Channel mode adaptation dynamically selects the type of traffic channel that a connection should be assigned to: either a full-rate (TCH/F) or a half-rate traffic channel (TCH/H). The basic idea here is to adapt a user's gross bit rate in order to optimize the use of radio resources. If the traffic load in a cell is high, those connections using a TCH/F (gross bit rate 22.8 kbit/s) and having high channel quality should be switched to a TCH/H (11.4 kbit/s). On the other hand, if the load is low, the speech quality of several TCH/H connections can be improved by switching them to a TCH/F. The signaling information for this type of adaptation is done with existing protocols on GSM signaling channels; the switching between full-rate and half-rate channels is realized by an intracell handover.

The task of codec mode adaptation is to adapt the coding rate (i.e. the tradeoff between the level of error protection versus the source bit rate) according to the current channel conditions. When the radio channel is bad, the encoder operates at low source bit rates at its input and uses more bits for forward error protection. When the quality of the channel is good, less error protection is employed.

The AMR codec consists of eight different modes with source/information bit rates ranging from 12.2 to 4.75 kbit/s (see Table 4.18). All modes are scaled versions of a common ACELP basis codec.

From the results of link quality measures, an adaptation unit selects the most appropriate codec mode. Figure 4.43 illustrates the AMR encoding principle. Channel coding is performed using a punctured recursive systematic convolutional code. Since not all bits of the voice data are equally important for audibility, AMR also employs an UEP structure. The most important bits (class Ia; e.g. mode bits and LPC coefficients) are additionally protected by a CRC code with six parity bits. On the receiver side, the decoder will discard the entire speech frame if the parity check fails. Also the degree of puncturing depends on the importance of the bits. At the end of the encoding process, a block with a fixed number of gross bits results, which is subsequently interleaved to reduce the number of burst errors.

Since the channel conditions can change rapidly, codec mode adaptation requires a fast signaling mechanism. This is achieved by transmitting the information about the used codec

Table 4.18 AMR codec modes.

	Source data rate (kbit/s)							
	12.2	10.2	7.95	7.4	6.7	5.9	5.15	4.75
Information bits per block	244	204	159	148	134	118	103	95
Class Ia bits (CRC protected)	81	65	75	61	55	55	49	39
Class Ib bits (not CRC protected)	163	139	84	87	79	63	54	56
Rate R of convolutional encoder	1/2	1/3	1/3	1/3	1/4	1/4	1/5	1/5
Output bits from convolutional encoder	508	642	513	474	576	520	565	535
Punctured bits	60	194	65	26	128	72	117	87

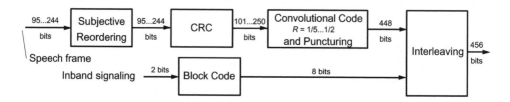

Figure 4.43 AMR channel encoding principle (bit numbers for TCH/F).

mode, link control, and DTX, etc. together with the speech data in the TCH, i.e. a special inband signaling is employed.

We give an example: the 12.2 kbit/s codec for a TCH/F operates with 244 source bits (12.2 kbit/s $\times$ 20 ms), which are first rearranged to subjective importance. By adding six CRC bits for class Ia bits, we obtain 250 bits. The subsequent recursive convolutional encoder, defined by the two generators 1 and $G_1/G_0 = (d^4 + d^3 + d + 1)/(d^4 + d^3 + d)$, with rate $R \approx 1/2$, maps those bits to 508 bits. Next, 60 bits are punctured, which results in an output sequence of 448 bits. Together with the encoded inband signaling (8 bits) this block is interleaved and finally mapped to bursts. The resulting gross bit rate is thus 456 bits/20 ms $= 22.8$ kbit/s.

4.9 Power-up scenario

At this point, all of the functions, protocols and mechanisms of the GSM radio interface have been presented which are needed to illustrate a basic power-up scenario. The following describes the basic events that occur during a power up of the MS. The scenario can be divided into several steps.

1. Provided that a SIM card is present, immediately after turning on power, a MS starts the search for BCCH carriers. Normally, the station has a stored list of up

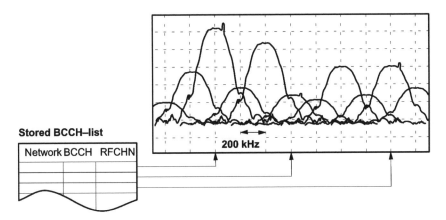

Figure 4.44 BCCH search in the power density spectrum (schematic).

to 32 carriers (Figure 4.44) of the current network. Signal level measurements are performed on each of these frequencies (RXLEV). Alternatively, if no list is available, all GSM frequencies have to be measured to find potential BCCH carriers. Using the path loss criterion C1 and the threshold values stored with the list of carriers (RXLEV_ACCESS_MIN, MS_TXPWR_MAX_CCH), a first ordering can be done.

2. After having found potential candidates based on the received signal level RXLEV, each carrier is investigated for the presence of an FCCH signal, beginning with the strongest signal. Its presence identifies the carrier as a BCCH carrier for synchronization. Using the sine wave signal allows coarse time synchronization as well as fine tuning of the oscillator.

3. The SB of the SCH in the TDMA frame immediately following the FCCH burst (Figure 4.17) has a long training sequence of 64 bits (Figure 4.7) which is used for fine tuning of the frequency correction and time synchronization. In this way the MS is able to read and decode synchronization data from the SCH, the BSIC and the RFN. This process starts with the strongest of all BCCH carriers. If a cell is identified using BSIC and path loss criterion C1, the cell is selected for camping on it.

4. The exact channel configuration of the selected cell is obtained from the BCCH data as well as the frequencies of the neighboring cells. The MS can now monitor the PCH of the current cell and measure the signal levels of the neighboring cells.

5. The MS must now prepare synchronization with the six cells with the strongest signal level (RXLEV) and read out their BCCH/SCH information, i.e. steps 1–4 above are to be performed continuously for the six neighboring cells with the best RXLEV values.

6. If significant changes are noticed using the path loss criterion C1 and the reselection criterion C2, the MS can start reselection of a new cell. Both criteria are determined periodically for the current BCCH and the six strongest neighbors.

To limit power consumption and to extend standby time of the battery, the mobile station can activate the DRX mode.

5

Protocols

5.1 Protocol architecture planes

The various physical aspects of radio transmission across the GSM air interface and the realization of physical and logical channels were explained in Chapter 4. According to the terminology of the OSI Reference Model, these logical channels are at the Service Access Point of Layer 1 (physical layer), where they are visible to the upper layers as transmission channels of the physical layer. The physical layer also includes the forward error correction and the encryption of user data.

The separation of logical channels into the two categories of control channels (signaling channels) and traffic channels (Table 4.1) corresponds to the distinction made in the ISDN Reference Model between user plane and control plane. Figure 5.1 shows a simplified reference model for the GSM User–Network Interface (UNI) Um, where the layer-transcending management plane is not elaborated on further in the following. In the user plane, protocols of the seven OSI layers are defined for the transport of data from a subscriber or a data terminal. User data are transmitted in GSM across the air interface over traffic channels TCH, which therefore belong to Layer 1 of the user plane (Figure 5.1).

Protocols in the signaling plane are used to handle subscriber access to the network and for the control of the user plane (reservation, activation, routing, switching of channels and connections). In addition, signaling protocols between network nodes are needed (network internal signaling). The Dm channels of the air interface in GSM are signaling channels and are therefore realized in the signaling plane (Figure 5.1).

Since signaling channels are physically present but mostly unused during an active user connection, it is obvious to use them also for the transmission of certain user data. In ISDN, packet-switched data communication is therefore permitted on the D channel, i.e. the physical D channel carries multiplexed traffic of signaling data (s-data) and user (payload) data (p-data). The same possibility also exists in GSM. Data transmission without allocation of a dedicated traffic channel is used for SMS by using free capacities on signaling channels. For this purpose, a separate SDCCH is allocated or, if a traffic connection exists, the SMS protocol data units are multiplexed onto the signaling data stream of the SACCH (Figure 5.2).

The control (signaling) and user plane can be defined and implemented separately, ignoring for the moment that control and user data have to be transmitted across the same

GSM – Architecture, Protocols and Services Third Edition   J. Eberspächer, H.-J. Vögel, C. Bettstetter and C. Hartmann
© 2009 John Wiley & Sons, Ltd

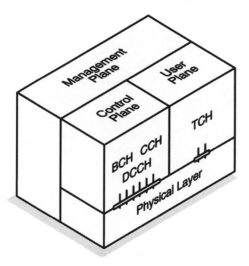

Figure 5.1 Logical channels at the air interface in the ISDN reference model.

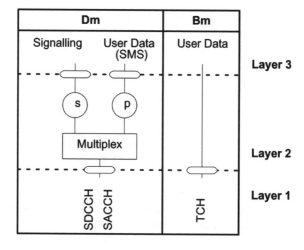

Figure 5.2 User data and control at the air interface.

physical medium at the air interface and that signaling procedures initiate and control activities in the user plane. Therefore, for each plane there exists a corresponding separate protocol architecture within the GSM system: the user data protocol architecture (section 5.2) and the signaling protocol architecture (section 5.3), with an additional separate protocol architecture for the transmission of p-data on the control (signaling) plane (section 5.3.2). A protocol architecture comprises not only the protocol entities at the radio interface Um but all protocol entities of the GSM network components.

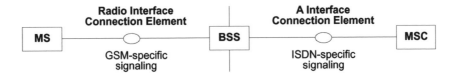

Figure 5.3 Connection elements.

5.2 Protocol architecture of the user plane

A GSM PLMN can be defined by a set of access interfaces (Appendix A.1) and a set of connection types used to realize the various communication services. A connection in GSM is defined between reference points. Connections are constructed from connection elements (Figure 5.3), and the signaling and transmission systems may change from element to element. Two elements therefore exist within a GSM connection: the radio interface connection element and the A interface connection element. The radio interface and the pertinent connection element are defined between the MS and the BSS, whereas the A interface connection element exists between BSS and MSC across the A interface. A GSM-specific signaling system is used at the radio interface, whereas ISDN-compatible signaling and payload transport are used across the A interface. The BSS is subdivided into BTS and BSC. Between them they define the Abis interface, which has no connection element defined; this is because it is usually transparent for user data.

A GSM connection type provides a way to describe GSM connections. Connection types represent the capabilities of the lower layers of the GSM PLMN. In the following section, the protocol models are presented as the basis for some of the connection types defined in the GSM standards. These are speech connections and transparent as well as nontransparent data connections. A detailed discussion of the individual connection types can be found in Appendix A with a description of how classical data services have been realized in GSM.

5.2.1 Speech transmission

The digital, source-coded speech signal of the MS is transmitted across the air interface in an error-protected and encrypted form. The signal is then deciphered in the BTS, and the error protection is removed before the signal is passed on. This specially protected speech transmission occurs transparently between the MS and a Transcoding and Rate Adaptation Unit (TRAU) which serves to transform the GSM speech-coded signals into the ISDN standard format (ITU-T A-law). A possible transport path for speech signals is shown in Figure 5.4, where the bit transport plane (encryption and TDMA/FDMA) has been omitted.

A simple GSM speech terminal (MT0, see also Figure A.1) contains a GSM Speech Codec (GSC) for speech coding. Its speech signals are transmitted to the BTS after channel coding (FEC) and encryption, where they are again deciphered, decoded and, if necessary, error-corrected. More than one GSM speech signal can be multiplexed onto an ISDN channel, with up to four GSM speech signals (at 13 kbit/s each) per ISDN B channel (64 kbit/s). Before they are passed to the MSC, speech signals are transcoded in the BSS from GSM format to ISDN format (ITU-T A-law).

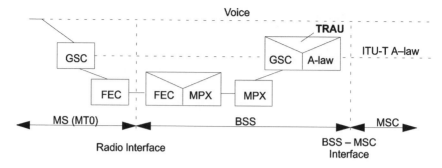

Figure 5.4 Speech transmission in GSM.

The BTSs are connected to the BSC over digital fixed lines, usually leased lines or microwave links, with typical transmission rates of 2048 kbit/s (in Europe), 1544 kbit/s (in the USA) or 64 kbit/s (ITU-T G.703, G.705, G.732). For speech transmission, the BSS implements channels of 64 or 16 kbit/s. The physical placement of the TRAU largely determines which kind of speech channel is used in the fixed network. The TRAU performs the conversion of speech data between GSM format (13 kbit/s) and ISDN A-law format (64 kbit/s). In addition, it is responsible for the adaptation of data rates, if necessary, for data services. There are two alternatives for the positioning of the TRAU: the TRAU can be placed in the BTS or outside the BTS in the BSC. An advantage of placing the TRAU outside the BTS is that up to four speech signals can be submultiplexed (MPX in Figure 5.4) onto an ISDN B channel, so that less bandwidth is required on the BTS-to-BSC connection. Beyond this consideration, placing the TRAU outside the BTS allows the TRAU functions to be combined for all BTSs of a BSS in one separate hardware unit, perhaps produced by a separate manufacturer. The TRAU is, however, always considered as part of the BSS and not as an independent network element.

Figure 5.5 shows some variants of TRAU placement. A BTS consists of a Base Control Function (BCF) for general control functions like frequency hopping, and several (at least one) Transceiver Function (TRX) modules which realize the eight physical TDMA channels on each frequency carrier. The TRX modules are also responsible for channel coding and decoding as well as encryption of speech and data signals. If the TRAU is integrated into the BTS, speech transcoding between GSM and ISDN formats is also performed within the BTS.

In the first case, TRAU within the BTS (BTS 1,2,3 in Figure 5.5), the speech signal in the BTS is transcoded into a 64 kbit/s A-law signal, and a single speech signal per B channel (64 kbit/s) is transmitted to the BSC/MSC. For data signals, the bit rates are adapted to 64 kbit/s, or several data channels are submultiplexed over one ISDN channel. The resulting user plane protocol architecture for speech transport is shown in Figure 5.6.

GSM-coded speech (13 kbit/s) is transmitted over the radio interface (Um) in a format that is coded for error protection and encryption. At the BTS site, the GSM signal is transcoded into an ISDN speech signal and transmitted transparently through the ISDN access network of the MSC.

In the second case, the TRAU resides outside the BTS (BTS 4 in Figure 5.5) and is considered as part of the BSC. However, physically it could also be located at the MSC site,

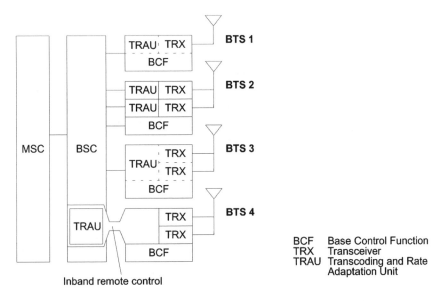

Figure 5.5 BTS architecture variations and TRAU placement.

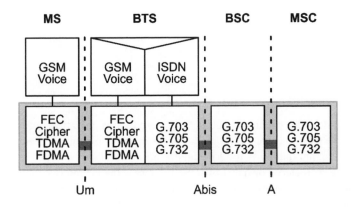

Figure 5.6 GSM protocol architecture for speech (TRAU at BTS site) Um.

i.e. at the MSC side of the BSC-to-MSC links (Figure 5.7). Channel coding/decoding and encryption are still performed in the TRX module of the BTS, whereas speech transcoding takes place in the BSC. For control purposes, the TRAU needs to receive synchronization and decoding information from the BTS, e.g. BFI for error concealment (section 4.7). If the TRAU does not reside in the BTS, it must be remotely controlled from the BTS by inband signaling. For this purpose, a subchannel of 16 kbit/s is reserved for the GSM speech signal on the BTS-to-BSC link, so an additional 3 kbit/s is made available for inband signaling. Alternatively, the GSM speech signal with added inband signaling could also be transmitted in a full ISDN B channel.

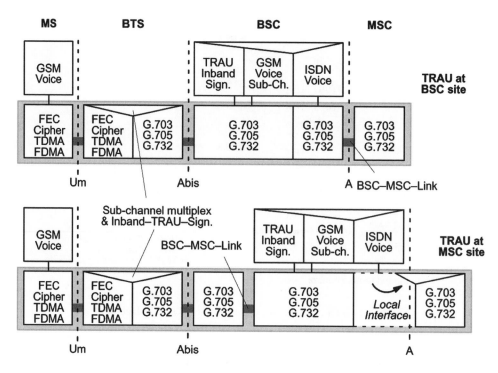

Figure 5.7 GSM protocol architecture for speech.

5.2.2 Transparent data transmission

The digital mobile radio channel is subject to severe quality variations and generates burst errors, which one tries to correct through interleaving and convolutional codes (section 4.8). However, if the signal quality is too low due to fading breaks or interference, the resulting errors cannot be corrected. For data transmission across the air interface Um, a residual bit error ratio varying between 10^{-2} and 10^{-5} according to channel conditions can be observed (Vögel *et al.*, 1995). This kind of variable quality of data transmission at the air interface determines the service quality of transparent data transmission. Transparent data transmission defines a GSM connection type used for the realization of some basic bearer services. The pertinent protocol architecture is illustrated in Figure 5.8. The main aspect of the transparent connection type is that user data are protected against transmission errors by forward error correction only across the air interface. Further transmission within the GSM network to the next MSC with an IWF to an ISDN or a Public Switched Telephone Network (PSTN) occurs unprotected on digital line segments, which already have a very low bit error ratio in comparison with the radio channel. The transparent GSM data service offers a constant throughput rate and constant transmission delay; however, the residual error ratio varies with channel quality due to the limited correction capabilities of the forward error correction.

For example, take a data terminal communicating over a serial interface of type V.24. A transparent bearer service provides access to the GSM network directly at a MS or through

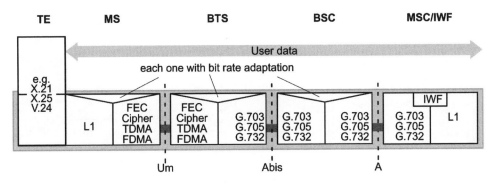

Figure 5.8 GSM protocol architecture for transparent data.

a terminal adapter (reference point R in Figure A.1). A data rate of up to 9.6 kbit/s can be offered based on the transmission capacity of the air interface and using an appropriate bit rate adaptation. The bit rate adaptation also performs the required asynchronous-to-synchronous conversion at the same time. This involves supplementing the tokens arriving asynchronously from the serial interface with fill data, since the channel coder requires a fixed block rate. In this way there is a digital synchronous circuit-switched connection between the terminal accessing the service and the IWF in the MSC, which extends across the air interface and the digital ISDN B channel inside the GSM network; this synchronous connection is completely transparent for the asynchronous user data of the Terminal Equipment (TE).

5.2.3   Nontransparent data transmission

Compared with the bit error ratio of the fixed network, which is of the order of 10^{-6} to 10^{-9}, the quality of the transparent data service is often insufficient for many applications, especially under adverse conditions. To provide more protection against transmission errors, more redundancy has to be added to the data stream. Since this redundancy is not always required, but only when there are residual errors in the data stream, forward error correction is inappropriate. Rather, an error-detection scheme with automatic retransmission of faulty blocks is used, Automatic Repeat Request (ARQ). Such an ARQ scheme which was specifically adapted to the GSM channel, is the RLP. The assumption for RLP is that the underlying forward error correction of the convolutional code realizes a channel with an average block error ratio of less than 10%, with a block corresponding to an RLP protocol frame of length 240 bits. Now the nontransparent channel experiences a constantly lower bit error ratio than the transparent channel, independent of the varying transmission quality of the radio channel; however, due to the RLP-ARQ procedure both throughput and transmission delay vary with the radio channel quality.

The data transmission between MS and interworking function of the next MSC is protected with the data link layer protocol RLP, i.e. the endpoints of RLP terminate in MS and IWF entities, respectively (Figure 5.9). At the interface to the data terminal TE, a Nontransparent Protocol (NTP) and an Interface Protocol (IFP) are defined, depending on the nature of the data terminal interface. Typically, a V.24 interface is used to carry character-oriented user data. These characters of the NTP are buffered and combined into blocks in the

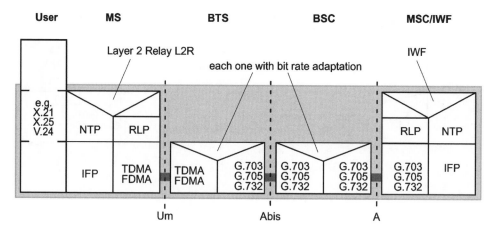

Figure 5.9 GSM protocol architecture for nontransparent data.

Layer 2 Relay (L2R) protocol, which transmits them as RLP frames. The data transport to and from the data terminal is flow-controlled. Therefore, transmission within the PLMN is no longer transparent for the data terminal. At the air interface, a new RLP frame is transmitted every 20 ms; thus L2R may have to insert fill tokens, if a frame cannot be completely filled at transmission time.

The RLP is very similar to the High-level Data Link Control (HDLC) protocol of ISDN (ISO/IEC 33091991) with regard to frame structure and protocol procedures, the main difference being the fixed frame length of 240 bits, in contrast to the variable length of HDLC. The frame consists of a 16-bit protocol header, 200-bit information field, and a 24-bit Frame Check Sequence (FCS); Figure 5.10. Owing to the fixed frame length, the RLP has no reserved flag pattern, and a special procedure to realize code transparency such as bit stuffing in HDLC is not needed. The very short – and hence less error prone – frames are exactly aligned with channel coding blocks. (The probability of frame errors increases with the length of the frame.)

RLP makes use of the services of the lower layers to transport its Protocol Data Units (PDUs). The channel offered to RLP therefore has the main characteristic of a 200 ms transmission delay, in addition to the possibly occurring residual bit errors. The delay is mostly caused by interleaving and channel coding, since the transmission itself takes only about 25 ms for a data rate of 9.6 kbit/s. This means it will take at least 400 ms until a positive acknowledgement is received for an RLP frame, and protocol parameters such as the transmission window and repeat timers need to be adjusted accordingly.

The RLP header is similar to that used in HDLC, with the difference that the RLP header contains no address information but only control information for which 16 bits are available. One distinguishes between supervisory frames and information frames. Whereas information frames carry user data, supervisory frames serve to control the connection (initialize, disconnect, reset) as well as the retransmission of information frames during data transfer. The information frames are labeled with a sequence number N(S) for identification, for which 6 bits are available in the RLP header (Figure 5.10). To conserve space, this

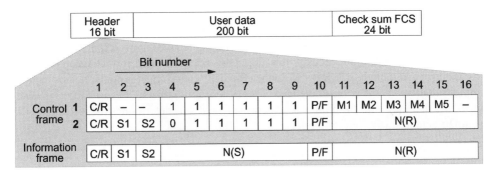

Figure 5.10 Frame structure of the RLP.

field is also used to code the frame type. Sequence number values smaller than 62 indicate that the frame carries user data in the information field (information frame). Otherwise the information field is discarded, and only the control information in the header is of interest (supervisory frame). These frames are marked with the reserved values 62 and 63 (Figure 5.10).

Owing to this header format, information frames can also carry (implicit) control information, a process known as piggybacking. The header information of the second variant can be carried completely within the header of an information frame. This illustrates further how RLP has been adapted to the radio channel, since it makes the transmission of additional control frames unnecessary during information transfer, which reduces the protocol overhead and increases the throughput.

Thus, the send sequence number is calculated modulo 62, which amounts to a window of 61 frames, allowing 61 outstanding frames without acknowledgement before the sender has to receive the acknowledgement of the first frame. Positive acknowledgement is used, i.e. the receiver sends an explicit supervisory frame as a receipt or an implicit receipt within an information frame. Such an acknowledgement frame contains a receive frame number N(R) which designates correct reception of all frames, including send sequence number N(S) = N(R) − 1.

Each time the last information frame is sent, a timer T1 is started at the sender. If an acknowledgement for some or all sent frames is not received in time, perhaps because the acknowledging RLP frame had errors and was therefore discarded, the timer expires and causes the sender to request an explicit acknowledgement. Such a request may be repeated N2 times; if this still leads to no acknowledgement, the connection is terminated. If an acknowledgement N(R) is obtained after expiration of timer T1, all sent frames starting from and including N(R) are retransmitted. In the case of an explicitly requested acknowledgement, this corresponds to a modified Go-back-N procedure. Such a retransmission is also allowed only up to N2 times. If no receipt can be obtained even after N2 trials, the RLP connection is reset or terminated.

Two procedures are provided in RLP for dealing with faulty frames: selective reject, which selects a single information frame without acknowledgement; and reject, which causes retransmission with implicit acknowledgement. With selective reject, the receiving RLP entity requests retransmission of a faulty frame with sequence number N(R), but this does not acknowledge receipt of other frames. Each RLP implementation must at least include the reject method for requesting retransmission of faulty frames. With a reject, the receiver asks for retransmission of all frames starting with the first defective received frame with number N(R) (Go-back-N). Simultaneously, this implicitly acknowledges correct reception of all frames up to and including N(R) − 1. Realization of selective reject is not mandatory in RLP implementations, but it is recommended. This is because Go-back-N causes retransmission of frames that may have been transmitted correctly and thus deteriorates the throughput that could be achieved with selective reject.

5.3 Protocol architecture of the signaling plane

5.3.1 Overview of the signaling architecture

Figure 5.11 shows the essential protocol entities of the GSM signaling architecture (control plane or signaling plane). Three connection elements are distinguished: the radio-interface connection element, the BSS-interface connection element and the A-interface connection element. This control plane protocol architecture consists of a GSM-specific part with the interfaces Um and Abis and a part based on SS#7 with the interfaces A, B, C, E (Figure 5.11). This change of signaling system corresponds to the change from radio interface connection element to A-interface connection element as discussed above for the user data plane (Figure 5.3).

The radio interface Um is defined between MS and BSS, more exactly between MS and BTS. Within the BSS, the BTS and the BSC cooperate over the Abis interface, whereas the A interface is located between BSC and MSC. The MSC has also signaling interfaces to VLR (B), HLR (C), to other MSCs (E) and to the EIR (F). Further signaling interfaces are defined between VLRs (G) and between VLR and HLR (D). Figure 3.3 gives an overview of the interfaces in a GSM PLMN.

Physical layer

In the control plane, the lowest layer of the protocol model at the air interface, the physical layer, implements the logical signaling channels (TDMA/FDMA, multiframes, channel coding, etc.; sections 4.7, 4.8 and 5.6). Like user data, signaling messages are transported over the Abis interface (BTS–BSC) and the A-interface (BSC–MSC) on digital lines with data rates of 2048 kbit/s (1544 kbit/s in the USA) or 64 kbit/s (ITU-T G.703, G.705, G.732).

Layer 2: LAPDm

On Layer 2 of the logical signaling channels across the air interface a data link protocol entity is implemented, the Link Access Procedure on Dm Channels (LAPDm). LAPDm is a derivative of Link Access Protocol Channel D (LAPD) which is specifically adapted to the air interface. This data link protocol is responsible for the protected transfer of signaling messages between MS and BTS over the air interface, i.e. LAPDm is terminated in MS and base station.

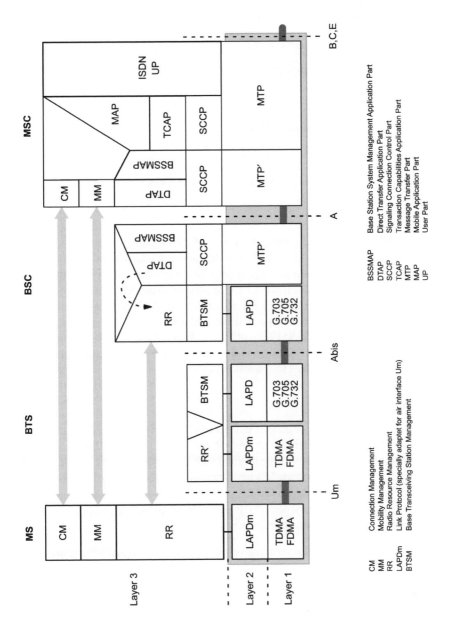

Figure 5.11 GSM protocol architecture for signaling.

In essence, LAPDm is a protocol similar to HDLC which offers a number of services on the various logical Dm channels of Layer 3: connection setup and teardown protected signaling data transfer. It is based on various link protocols used in fixed networks, such as LAPD in ISDN (Bocker, 1990). The main task of LAPDm is the transparent transport of messages between protocol entities of Layer 3 with special support for:

- multiple entities in Layer 3 and Layer 2;

- signaling for broadcasting (BCCH);

- signaling for paging (PCH);

- signaling for channel assignment (AGCH);

- signaling on dedicated channels (SDCCH).

A detailed discussion of LAPDm is presented in section 5.4.2.

Layer 3

In the MS, the LAPDm services are used at Layer 3 of the signaling protocol architecture. Layer 3 is divided into three sublayers: Radio Resource Management (RR), Mobility Management (MM) and Connection Management (CM). The protocol architecture formed by these three sublayers is shown in Figure 5.12. CM is further subdivided into three protocol entities: Call Control (CC), Supplementary Services (SS) and SMS. Additional multiplexing functions within Layer 3 are required between these sublayers.

The call-independent SS and the SMS are offered to higher layers at two Service Access Points (SAPs), MNSS and MNSMS. A more detailed look at the services offered by the RR, MM and CC protocol entities is given in the following.

Radio resource management

RR essentially handles the administration of the frequencies and channels. This involves the RR module of the MS communicating with the RR module of the BSC (Figure 5.11). The general objective of the RR is to set up, maintain and take down RR connections which enable point-to-point communication between MS and network. This also includes cell selection in idle mode and handover procedures. Furthermore, the RR is responsible for monitoring BCCH and CCCH on the downlink when no RR connections are active.

The following functions are realized in the RR module:

- monitoring of BCCH and PCH (readout of system information and paging messages);

- RACH administration: MSs send their requests for connections and replies to paging announcements to the BSS;

- requests for and assignments of data and signaling channels;

- periodic measurement of channel quality (quality monitoring);

- transmitter power control and synchronization of the MS;

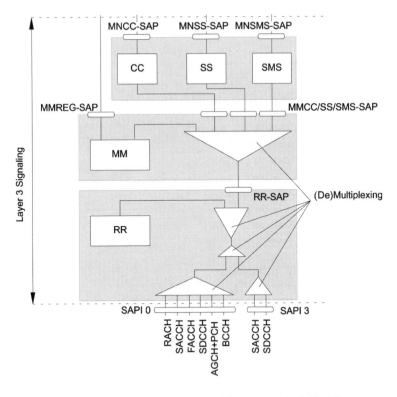

Figure 5.12 Layer 3 protocol architecture at the MS side.

- handover (part of which is sometimes erroneously attributed to roaming functions and MM), always initiated by the network;

- synchronization of encryption and decryption on the data channel.

The RR sublayer provides several services at the RR-SAP to the MM sublayer. These services are needed to set up and take down signaling connections and to transmit signaling messages.

Mobility management

MM encompasses all of the tasks resulting from mobility. The MM activities are exclusively performed in cooperation between MS and MSC, and they include:

- TMSI assignment;

- localization of the MS;

- location updating of the MS, sometimes known as roaming functions;

- identification of the MS (IMSI, IMEI);

Table 5.1 Distribution of functions between BTS, BSC and MSC (according to GSM Rec. 08.02, 08.52).

	BTS	BSC	MSC
Terrestrial channel management			
MSC-BSC-channels			
Channel allocation			X
Blocking indication		X	
BSC-BTS-Channels			
Channel allocation		X	
Blocking indication	X		
Mobility management			
Authentication			X
Location updating			X
Call control			X
Radio channel management			
Channel coding/decoding	X		
Transcoding/rate adaptation	X		
Interworking function			X
Measurements			
Uplink measuring	X		X
Processing of reports from MS/TRX	X	X	X
Traffic measurements			X
Handover			
BSC internal, intracell		X	
BSC internal, intercell		X	
BSC external		X	
Recognition, decision, execution			X
HO access detection	X		
Paging			
Initiation		X	
Execution	X		
Channel configuration management		X	
Frequency hopping			
Management		X	
Execution	X		
TCH management			
Channel allocation		X	
Link supervision		X	
Channel release		X	X
Idle channel observation	X		
Power control determination	X	X	

Table 5.1 Continued.

	BTS	BSC	MSC
SDCCH management			
SDCCH allocation		X	
Link supervision		X	
Channel release		X	X
Power control determination	X	X	
BCCH/CCCH management			
Message scheduling management		X	
Message scheduling execution	X		
Random access detection	X		
Immediate assign		X	
Timing Advance			
Calculation	X		
Signaling to MS at random access		X	
Signaling to MS at handover/during call	X		
Radio resource indication			
Report status of idle channels	X		
LAPDm functions	X		
Encryption			
Management		X	
Execution	X		

- authentication of the MS;

- IMSI attach and detach procedures (e.g. at insertion or removal of SIM);

- ensuring confidentiality of subscriber identity.

Registration services for higher layers are provided by Layer 3 at the MMREG-SAP (Figure 5.12). Registration involves the IMSI attach and detach procedures which are used by the mobile to report state changes such as power-up or power-down, or SIM card removal or insertion.

The MM sublayer offers its services at the MMCC-SAP, MMSS-SAP and MMSMS-SAP to the CC, SS and SMS entities. This is essentially a connection to the network side over which these units can communicate.

Connection management

CM consists of three entities: CC, SS and SMS. CC handles all tasks related to setting up, maintaining and taking down calls. The services of CC are provided at the MNCC-SAP, and they encompass:

- establishment of normal calls (MS-originating and MS-terminating);

- establishment of emergency calls (only MS-originating);

- termination of calls;

- Dual-Tone Multifrequency (DTMF) signaling;

- call-related supplementary services;

- in-call modification: the service may be changed during a connection (e.g. speech and transparent/nontransparent data are alternating; or speech and fax alternate).

The service primitives at this SAP of the interface to higher layers report reception of incoming messages and effect the sending of messages, essentially ISDN user–network signaling according to Q.931.

RR messages are mainly exchanged between MS and BSS. In contrast, CM and MM functions are handled exclusively between MS and MSC; the exact division of labor between BTS, BSC and MSC is summarized in Table 5.1. As can be seen, RR messages have to be transported over the Um and Abis interfaces, whereas CM and MM messages need additional transport mechanisms across the A interface.

Message transfer part

From a conceptual viewpoint, the A interface in GSM networks is the interface between the MSCs, the ISDN exchanges with mobile network specific extensions, and the BSC, the dedicated mobile network specific control units. Here too is the reference point, where the signaling system changes from GSM-specific to the general ISDN-compatible SS#7. Message transport in the SS#7 network is realized through the Message Transfer Part (MTP). In essence, MTP comprises the lower three layers of the OSI Reference Model, i.e. the MTP provides routing and transport of signaling messages.

A slightly modified (reduced) version of the MTP, called MTP', has been defined for the protected transport of signaling messages across the A interface between BSC and MSC. At the ISDN side of the MSCs, the complete MTP is available. For signaling transactions between MSC and MS (CM, MM), it is necessary to establish and identify distinct logical connections. The Signaling Connection Control Part (SCCP) is used for this purpose to facilitate implementation with a slightly reduced range of functions defined in SS#7.

BSS application part

For GSM-specific signaling between MSC and BSC, the Base Station System Application Part (BSSAP) has been defined. The BSSAP consists of the Direct Transfer Application Part (DTAP) and the Base Station System Management Application Part (BSSMAP). The DTAP is used to transport messages between MSC and MS. These are the CC and MM messages. At the A interface, they are transmitted with DTAP and then passed transparently through the BSS across the Abis interface to the MS without interpretation by the BTS.

The BSSMAP is the protocol definition part which is responsible for all of the administration and control of the radio resources of the BSS. RR is one of the main functions of a BSS. Therefore, the RR entities terminate in the MS and the BTS or BSC, respectively. Some functions of RR, however, require involvement of the MSC (e.g. some handover situations, or release of connections or channels). Such actions should be initiated and controlled by the MSC (e.g. handover and channel assignment). This control is the responsibility of BSSMAP.

RR messages are mapped and converted within the BSC into procedures and messages of BSSMAP and vice versa. BSSMAP offers the functions which are required at the A interface between BSS and MSC for RR of the BSS. Accordingly, RR messages initiate BSSMAP functions, and BSSMAP functions control RR protocol functions.

BTS management

A similar situation exists at the Abis interface. Most of the RR messages are passed transparently by the BTS between MS and BSC. Certain RR information, however, must be interpreted by the BTS, e.g. in situations such as random access of the MS, the start of the ciphering process, or paging to localize a MS for connection setup. The Base Transceiver Station Management (BTSM) contains functions for the treatment of these messages and other procedures for BTS management. In addition a mapping occurs in the BTS from BTSM onto the RR messages relevant at the air interface (RR', Figure 5.11).

Mobile application part

The MSC is equipped with the MAP, a mobile network specific extension of SS#7, for communication with the other components of the GSM network (the HLR and VLR registers, other MSCs) and other PLMNs. Among the MAP functions are all signaling functions among MSCs as well as between MSC and the registers (Figure 5.13). These functions include:

- updating of residence information in the VLR;

- cancellation of residence information in the VLR;

- storage of routing information in the HLR;

- updating and supplementing of user profiles in HLR and VLR;

- inquiry of routing information from the HLR;

- handover of connections between MSCs.

The exchange of MAP messages, e.g., with other MSCs, HLR or VLR, occurs over the transport and transaction protocol of the SS#7. The SS#7 transaction protocol is the Transaction Capabilities Application Part (TCAP). A connectionless transport service is offered by the SCCP.

The MAP functions require channels for signaling between different PLMNs which are provided by the international SS#7. Access to SS#7 occurs through the fixed ISDN.

Connection to the fixed network is typically done through leased lines; in the case of the German GSM network operators, it is through lines with a rate of 2 Mbit/s from Deutsche Telekom (Gottschalk, 1993). Often the majority of the MSC in a PLMN has such an access to the fixed network. On these lines, both user data and signaling data are transported. From the viewpoint of a fixed network, an MSC is integrated into the network in the same way as a normal ISDN exchange node. Outside of a PLMN, starting with the GMSC, calls for MSs are treated like calls for subscribers of the fixed network, i.e. the mobility of a subscriber with an MSISDN becomes 'visible' only beyond the GMSC. For CC, the MSC has the same interface as an ISDN switching node. Connection-oriented signaling of GSM networks is mapped at

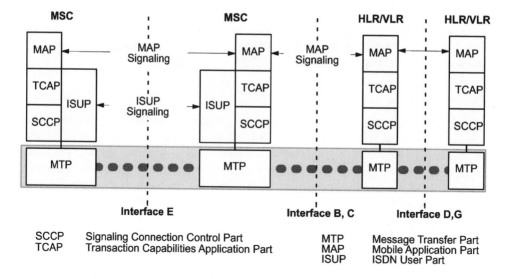

Figure 5.13 Protocol interfaces in the mobile network.

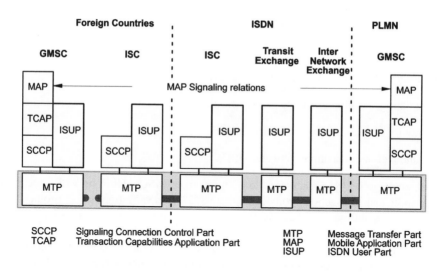

Figure 5.14 International signaling relations via ISDN (Gottschalk, 1993).

the fixed network side (interface to ISDN) into the ISDN User Part (ISUP) used to connect ISDN channels through the network (Figure 5.14). The mobile-specific signaling of the MAP is routed over a gateway of the PLMN (GMSC) and the ISC of the national ISDN network into the international SS#7 network (Gottschalk, 1993). In this way, transport of signaling data between different GSM networks is also guaranteed without problems.

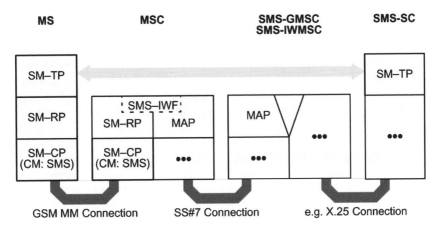

Figure 5.15 Protocol architecture for SMS transfer.

5.3.2   Transport of user data in the signaling plane

In the signaling plane (control plane) of the GSM architecture, one can also transport packet-oriented user data from or to MSs. This occurs for the point-to-point SMS. Short messages are always transmitted in store-and-forward mode through a Short Message Service Center (SMS-SC). The service center accepts these messages, which can be up to 160 characters long, and forwards them to the recipients (other MSs or fax, email, etc.). In principle, GSM defines a separate protocol architecture for the realization of this service.

Between MS and service center, short messages are transmitted using a connectionless transport protocol: Short Message Transport Protocol (SM-TP) which uses the services of the signaling protocols within the GSM network. Transport of these messages outside of the GSM network is not defined. For example, the SMS-SC could be directly connected to the gateway switching center (SMS-GMSC), or it could be connected to a Short Message Service Interworking MSC (SMS-IWMSC) through an X.25 connection (Figure 5.15). Within the GSM network between MSCs, a short message is transferred with the MAP and the lower layers of SS#7. Finally, between a MS and its local MSC, two protocol layers are responsible for the transfer of transport protocol units of SMS. First, there is the SMS entity in the CM sublayer of Layer 3 at the user–network interface (see Figure 5.12) which realizes the Short Message Control Protocol (SM-CP) and its connection-oriented service. Second, there is the relay layer, in which the Short Message Relay Protocol (SM-RP) is defined, which offers a connectionless service for transfer of SMS transport PDUs between MS and MSC. This, however, uses services at the service access point MMSMS-SAP (see Figure 5.12) and thus a connection of the MM sublayer.

In addition to the SM-CP, the relay protocol SM-RP was introduced above the CM sublayer (Figures 5.12 and 5.15) to realize an acknowledged transmission of short messages, but with minimal overhead for the radio channel. A short message sent by a MS is passed over the signaling network until it reaches the service center SMS-SC. If the service center determines the error-free reception of a message, an acknowledgement message is returned on the reverse path, which finally causes sending of an acknowledgement message from the SM-RP entity in the MSC to the MS. Until this acknowledgement message arrives, the connection in the MM sublayer can be taken down, and thus also the reserved radio channel. In this way, radio resources across the air interface are only occupied during the actual transmission of SM-RP messages. Each successful transmission of an SM-CP PDU across the MM connection, which includes the error-prone air interface, is immediately acknowledged, or else errors are immediately reported to the sending SM-CP entity. So if a message is damaged at the radio interface, this avoids it being transmitted to the service center.

5.4 Signaling at the air interface (Um)

Signaling at the user–network interface in GSM is essentially concentrated in Layer 3. Layers 1 and 2 provide the mechanisms for the protected transmission of signaling messages across the air interface. In addition to the local interface, they contain functionality and procedures for the interface to the BTS. The signaling of Layer 3 at the user–network interface is very complex and comprises protocol entities in the MS and in all functional entities of the GSM network (BTS, BSC and MSC).

5.4.1 Layer 1 of the MS-BTS interface

Layer 1 of the OSI Reference Model (physical layer) contains all of the functions necessary for the transmission of bit streams over the physical medium, in this case the radio channel. GSM Layer 1 defines a series of logical channels based on the channel access procedures with their physical channels. The higher layer protocols access these services at the Layer 1 service interface. The three interfaces of Layer 1 are schematically illustrated in Figure 5.16.

LAPDm protocol frames are transmitted across the service mechanisms of the data link layer interface, and the establishment of logical channels is reported to Layer 2. The communication across this interface is defined by abstract physical layer service primitives. A separate SAP is defined for each logical control channel (BCCH, PCH + AGCH, RACH, SDCCH, SACCH, FACCH).

There is a direct interface between Layer 1 and the RR sublayer of Layer 3. The abstract service primitives exchanged at this interface mostly concern channel assignment and Layer 1 system information, including measurement results of channel monitoring. At the third Layer 1 interface, the traffic channels for user (payload) data are provided.

The SAPs of Layer 1 as defined in GSM are not genuine SAPs in the spirit of OSI. They differ from the PHY-SAPs of the OSI Reference Model insofar as these SAPs are controlled by Layer 3 RR sublayer (layer management, establishment and release of channels) rather than by control procedures in the link layer. Control of Layer 1 SAPs by RR comprises activation and deactivation, configuration, routing and disconnection of physical and logical channels. Furthermore, exchange of measurement and control information for channel monitoring occurs through service primitives.

Layer 1 services

Layer 1 services of the GSM user–network interface are divided into three groups:

- access capabilities;

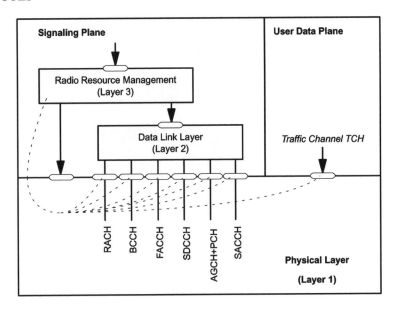

Figure 5.16 Layer 1 service interfaces.

- error detection;

- encryption.

Layer 1 provides a bit transport service for the logical channels. These are transmitted in multiplexed format over physical channels which consist of elements defined for the transmission on the radio channel (frequency, time slot, hopping sequence, etc.; section 5.1). Some physical channels are provided for common (shared) use (BCCH and CCCH), whereas others are assigned to dedicated connections with single MSs (dedicated physical channels). The combination of logical channels used on a physical channel can vary over time, e.g. TCH + SACCH/FACCH replaced by SDCCH + SACCH (see Table 4.4).

The GSM standard distinguishes explicitly between access capabilities for dedicated physical channels and for common physical channels BCCH/CCCHs. Dedicated physical channels are established and controlled by Layer 3 RR management. During the operation of a dedicated physical channel, Layer 1 continuously measures the signal quality of the used channel and the quality of the BCCH channels of the neighboring base stations. This measurement information is passed to Layer 3 in measurement service primitives Management Physical Headers (MPH). In idle mode, Layer 1 selects the cell with the best signal quality in cooperation with the RR sublayer based on the quality of the BCCH/CCCH (cell selection).

GSM Layer 1 offers an error-protected bit transport service and therefore also error detection and correction mechanisms. To do this, error-correcting and error-detecting coding mechanisms are provided (section 4.8). Frames recognized as faulty are not passed to Layer 2. Furthermore, security-relevant functions such as encryption of user data are implemented in Layer 1 (section 5.6).

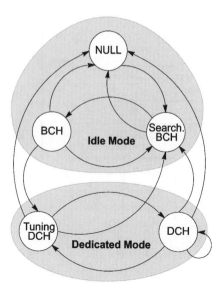

Figure 5.17 State diagram of the physical layer of MS.

Layer 1: Procedures and peer-to-peer signaling

GSM defines and distinguishes between two operational modes of a MS: idle mode and dedicated mode (Figure 5.17). In idle mode, the MS is either powered off (state NULL) or it searches for or measures the BCCH with the best signal quality (state SEARCHING BCH), or is synchronized to a specific base station's BCCH and ready to perform a random access procedure on the RACH for requesting a dedicated channel in state BCH (section 4.5.4).

In state TUNING DCH of the dedicated mode, the MS occupies a physical channel and tries to synchronize with it, which will eventually result in transition to state DCH. In this state, the MS is finally ready to establish logical channels and switch them through. The state transitions of Layer 1 are controlled by MPH service primitives of the RR interface, i.e. directly from the Layer 3 RR sublayer of the signaling protocol stack.

Layer 1 defines its own frame structure for the transport of signaling messages, which occur as LAPDm frames at the SAP of the respective logical channel. Figure 5.18 shows the format of an SACCH block as an example, which essentially contains 21 octets of LAPDm data.

Furthermore, the SACCH frame contains a kind of protocol header which carries the current power level and the value of the timing advance. This header is omitted in the other logical channels (FACCH, SDCCH, CCCH and BCCH) which contain only LAPDm PDUs.

5.4.2   Layer 2 signaling

The LAPDm protocol is the data link protocol for signaling channels at the air interface. It is similar to HDLC. It provides two operational modes:

- unacknowledged operation;
- acknowledged operation.

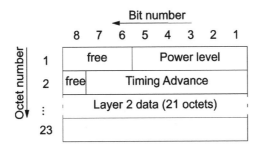

Figure 5.18 Format of an SACCH block.

In the unacknowledged operation mode, data is transmitted in Unnumbered Information (UI) frames without acknowledgement; there is no flow control or L2 error correction. This operational mode is allowed for all signaling channels, except for the RACH which is accessed in multiple access mode without reservation or protection.

The acknowledged operation mode provides protected data service. Data are transmitted in information (I) frames with positive acknowledgement. Error protection through retransmission (ARQ) and flow control are specified and activated in this mode. This mode is only used on DCCH channels.

In LAPDm, the Connection End Points (CEPs) of L2 connections are labeled with Data Link Connection Identifiers (DLCIs), which consist of two elements.

- The Layer 2 Service Access Point Identifier (SAPI) is transmitted in the header of the L2 protocol frame.

- The physical channel identifier on which the L2 connection is or will be established, is the real Layer 2 Connection End Point Identifier (CEPI). The CEPI is locally administered and not communicated to the L2 peer entity. (The terminology of the GSM standard is somewhat inconsistent in this case – what is really meant is the respective logical channel. The physical channels from the viewpoint of LAPDm are the logical channels of GSM, rather than the physical channels defined by frequency/time slot/hopping sequence.)

When a Layer 3 message is transmitted, the sending entity chooses the appropriate SAP and CEP. When the Service Data Unit (SDU) is handed over at the SAP, the chosen CEP is given to the L2 entity. Conversely, when receiving an L2 frame, the appropriate L2-CEPI can be determined from the physical/logical channel identity and the SAPI in the frame header.

Specific SAPI values are reserved for the certain functions:

- SAPI = 0 for signaling (CM, MM, RR);

- SAPI = 3 for SMS.

In the control plane, these two SAPI values serve to separate signaling messages from packet-oriented user data (short messages). Further functions needing a new SAPI value can be defined in future versions of the GSM standard.

Table 5.2 Logical channels, operational modes and Layer 2 SAPIs.

Logical channel	SAPI = 0	SAPI = 3
BCCH	Unacknowledged	–
CCCH	Unacknowledged	–
SDCCH	Unacknowledged and acknowledged	Unacknowledged and acknowledged
SACCH associated with SDCCH	Unacknowledged	–
SACCH associated with TCH	Unacknowledged	Unacknowledged and acknowledged
FACCH	Unacknowledged and acknowledged	–

An LAPDm entity is established for each of the pertinent physical/logical channels. For some of the channel/SAPI combinations only a subset of the LAPDm protocol is needed (e.g. unacknowledged operation), and some channel/SAPI combinations are not supported (Table 5.2). These LAPDm entities perform the data link procedure, i.e. the functions of the L2 peer-to-peer communication as well as the service primitives between adjacent layers. Segmentation and reassembly of Layer 3 messages is also included.

Further Layer 2 procedures are the distribution procedure and the Random Access (RA) procedure. The distribution procedure is needed if multiple SAPs are associated with one physical/logical channel. It performs the distribution of the L2 frames received on one channel to the respective data link procedure or the priority-controlled multiplexing of L2 frames from multiple SAPs onto one channel. The random access procedure is used on the RACH; it deals with the random controlled retransmission of random access bursts, but it does not perform any error protection on the unidirectional RACH.

For certain aspects of RR, the protocol logic of Layer 3 has to have direct access to the services of Layer 1. In particular, this is needed for the functions of radio subsystem link control, i.e. for channel measurement, transmitter power control and timing advance.

A possible link layer configuration of a MS is shown in Figure 5.19. The base station has a similar configuration with one PCH + AGCH, SDCCH and SACCH/FACCH for each active MS.

Figure 5.20 shows the different types of protocol data frames used for communication between L2 peer entities in MS and BTS. Frame formats A and B are used on the SACCH, FACCH and SDCCH channels, depending upon whether the frame has an information field (type B) or not (type A). For unacknowledged operation (BCCH, PCH, AGCH), format types Abis and Bbis are used on channels with SAPI = 0. The Abis format is used when there is no information to be transmitted on the respective logical channel.

In contrast to HDLC, LAPDm frames have no flag to designate the beginning and end of a frame, rather the delineation of frames at the link level is performed as in RLP (section 5.2.3) through the fixed-length block structure of Layer 1. The maximum number of octets N201 per information field depends on the type of logical channel (Table 5.3). The end of the information field is given by a length indicator, a value of less than N201 indicates that the frame has to be supplemented with fill bits to the full length. In the case of an SACCH

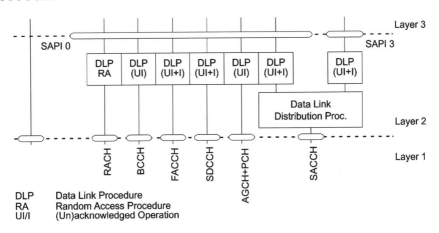

DLP   Data Link Procedure
RA   Random Access Procedure
UI/I   (Un)acknowledged Operation

Figure 5.19 Sample configuration of the MS data link layer.

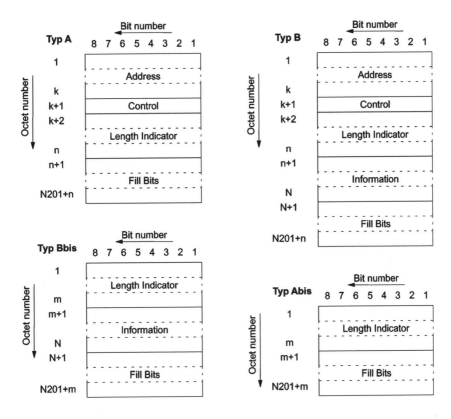

Figure 5.20 LAPDm frame formats.

Table 5.3 Logical channels and the maximum length of the LAPDm information field.

Logical channel	N201
SACCH	18 octets
SDCCH, FACCH	20 octets
BCCH, AGCH, PCH	22 octets

channel, for example, this yields a fixed-length LAPDm packet of 21 octets. Combined with the fields for transmitter power control and timing advance, an SACCH block of Layer 1 is thus 23 octets long.

The address field may have a variable length, however; for use on control channels it consists of exactly one octet. In addition to other fields, this octet contains a SAPI (3 bits) and the Command/Response (C/R) flag known from HDLC. In LAPDm, the coding of the control field with sending and receiving sequence numbers and the state diagram describing the protocol procedures are almost identical to HDLC (ISO/IEC 33091991, 2008). Some additional parameters are required at the service interface to Layer 3; for example, a parameter CEP designating the desired logical channel. Furthermore, the LAPDm protocol has some simplifications or peculiarities with regards to HDLC.

- The sending window size is restricted to $k = 1$.

- The protocol entities should be implemented in such a way that the state RECEIVER BUSY is never reached. Thus, RNR packets can be safely ignored. The HDLC polling procedure for state inquiry of the partner station need not be implemented in LAPDm.

- Connections to SAPI = 0 are always initiated by the MS.

In addition, the repetition timer T200 and the maximum number of allowed repetitions N200 have been adapted to the special needs of the mobile channel. In particular, they have their own value determined by the type of logical channel.

5.4.3 Radio resource management

The procedures for RR are the basic signaling and control procedures at the air interface. They handle the assignment, allocation and administration of radio resources, the acquisition of system information from broadcast channels (BCCH) and the selection of the cell with the best signal reception (see cell selection in section 4.5.4). Accordingly, the RR procedures and pertinent messages (Table 5.4) are defined for idle mode as well as for setting up, maintaining and taking down of RR connections.

Figure 5.21 shows the format of RR messages, which is uniform for all three Layer 3 signaling sublayers (CM, MM, RR). Each Layer 3 message contains a protocol discriminator in the first octet, which allows association of messages with the respective sublayer or service access point (Figure 5.12). The uppermost four bits of the first octet also contain a transaction ID, which enables a MS to perform several signaling transactions in parallel.

Table 5.4 RR messages.

Category	Message	Logical channel	Direction	MT code
Channel establishment	Additional assignment	DCCH	N → MS	00111011
	Immediate assignment	CCCH	N → MS	00111111
	Immediate assignment extended	CCCH	N → MS	00111001
	Immediate assignment rejected	CCCH	N → MS	00111010
Ciphering	Ciphering mode command	DCCH	N → MS	00110101
	Ciphering mode complete	DCCH	MS → N	00110010
Handover	Assignment command	DCCH	N → MS	00101110
	Assignment complete	DCCH	MS → N	00101001
	Assignment failure	DCCH	MS → N	00101111
	Handover access	DCCH	MS → N	–
	Handover command	DCCH	N → MS	00101011
	Handover complete	DCCH	MS → N	00101100
	Handover failure	DCCH	MS → N	00101000
	Physical information	DCCH	N → MS	00101101
Channel release	Channel release	DCCH	N → MS	00001101
	Partial release	DCCH	N → MS	00001010
	Partial release complete	DCCH	MS → N	00001111
Paging	Paging request, type 1/2/3	PCH	N → MS	00100xxx
	Paging response	DCCH	MS → N	00100111
System information	System information, type 1/2/3/4	BCCH	N → MS	00011xxx
	System information, type 5/6	SACCH	N → MS	00011xxx
Miscellaneous	Channel mode modify	DCCH	N → MS	00010000
	Channel mode modify acknowledge	DCCH	MS → N	00010111
	Channel request	RACH	MS → N	–
	Classmark change	DCCH	MS → N	00010110
	Frequency redefinition	DCCH	N → MS	00010100
	Measurement report	SACCH	MS → N	00010101
	Synchronization channel information	SCH	N → MS	–
	RR status	DCCH	MS ↔ N	00010010

The Message Type (MT) is registered in the lower seven bits of the second octet (Tables 5.4–5.6). Otherwise, Layer 3 messages consist of Information Elements (IEs) of fixed or variable length; a Length Indicator (LI) is added for variable-length messages.

In idle mode, the MS is reading the BCCH information continuously and conducts periodic measurements of the signaling strength of the BCCH carriers in order to be able to select the current cell (section 4.5.4). In this state, there is no exchange of signaling messages with the network. The data required for RR and other signaling procedures are collected and stored: the list of neighboring BCCH carriers, thresholds for RR algorithms, CCCH configurations, information about the use of RACH and PCH, etc. This information is broadcast by the BSS on the BCCH (SYSTEM INFORMATION, types 1–4) and therefore

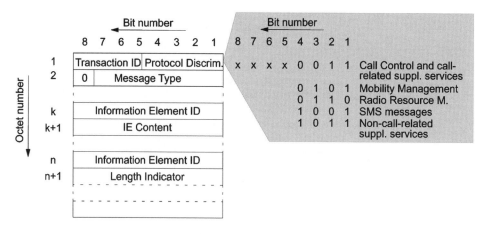

Figure 5.21 Format of a Um signaling message (Layer 3).

Table 5.5 MM messages.

Category	Message	Direction	MT
Registration	IMSI detach indication	MS → N	0x000001
	Location updating accept	N → MS	0x000010
	Location updating reject	N → MS	0x000100
	Location updating request	MS → N	0x001000
Security	Authentication reject	N → MS	0x010001
	Authentication request	N → MS	0x010010
	Authentication response	MS → N	0x010100
	Identity request	N → MS	0x001000
	Identity response	MS → N	0x001001
	TMSI reallocation command	N → MS	0x001010
	TMSI reallocation complete	MS → N	0x001011
Connection management	CM service accept	MS ↔ N	0x100001
	CM service reject	N → MS	0x100010
	CM service request	MS → N	0x100100
	CM reestablishment request	MS → N	0x101000
Miscellaneous	MM status	MS → N	0x110001

is available to all MSs currently in the cell. Also important is the periodic monitoring of the PCH so that paging calls are not lost. For this purpose, the BSS is sending on all paging channels of a cell continuously valid Layer 3 messages (PAGING REQUEST) which the MS can decode and recognize if its address is paged.

Table 5.6 CC messages for circuit-switched connections.

Category	Message	Direction	MT
Call establishment	Alerting	N → MS	0x000001
	Call confirmed	MS → N	0x001000
	Call proceeding	N → MS	0x000010
	Connect	N ↔ MS	0x000111
	Connect acknowledge	N ↔ MS	0x001111
	Emergency setup	MS → N	0x001110
	Progress	N → MS	0x000011
	Setup	N ↔ MS	0x000101
Call information phase	Modify	N ↔ MS	0x010111
	Modify complete	N ↔ MS	0x011111
	Modify reject	N ↔ MS	0x010011
	User information	N ↔ MS	0x010000
Call clearing	Disconnect	N ↔ MS	0x100101
	Release	N ↔ MS	0x101101
	Release complete	N ↔ MS	0x101010
Miscellaneous	Congestion control	N ↔ MS	0x111001
	Notify	N ↔ MS	0x111110
	Start DTMF	MS → N	0x110101
	Start DTMF acknowledge	N → MS	0x110010
	Start DTMF reject	N → MS	0x110111
	Status	N ↔ MS	0x111101
	Status enquiry	N ↔ MS	0x110100
	Stop DTMF	MS → N	0x110001
	Stop DTMF acknowledge	N → MS	0x110010

Connection setup and release

Each exchange of signaling messages with the network (BSS, MSC) requires a RR connection and the establishment of an LAPDm connection between MS and BTS. Setting up the RR connection can be initiated by the network or the MS (Figure 5.22). In either case, the MS sends a channel request (CHANQUEST) on the RACH in order to get a channel assigned on the AGCH (immediate assignment procedure). There is also a procedure to deny a channel request (immediate assignment reject).

If the network does not immediately answer the channel request, the request is repeated using the Aloha method with a random number controlled timer (Figure 5.22). In the case of a network-initiated connection, this procedure is preceded by a paging call (PAGING REQUEST) to be answered by the MS (PAGING RESPONSE). After a RR connection has been successfully completed, the higher protocol layers (CM, MM) can receive and transmit signaling messages at SAPI 0.

In contrast to the setup of connections, the release is always initiated by the network (CHANNEL RELEASE). Reasons for the release of the channel could be end of the signaling

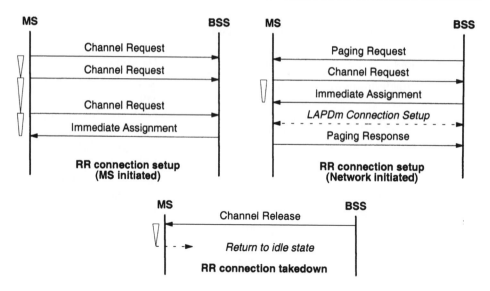

Figure 5.22 RR connection setup and takedown.

transaction, too many errors, removal of the channel in favor of a higher priority call (e.g. emergency call) or end of a call. After receiving the channel release command, the MS assumes the idle state following a brief waiting period (Figure 5.22).

Once a RR connection has been set up, the MS has either a SDCCH or a TCH with associated SACCH/FACCH available for exclusive bidirectional use. On the SACCH, data must be sent continuously (see also section 4.5.3), i.e. the MS keeps sending current channel measurements (MEASUREMENT REPORT; section 4.5.1) if no other signaling messages need to be sent. In the other direction, the BSS keeps sending system information (SYSTEM INFORMATION, alternating between type 5 and type 6). The information element with the coded measurement results contains the following among other data: RXLEV and RXQUAL of the serving cell as well as RXLEV and carrier frequency of up to six neighboring cells as well as their BSICs (Figure 5.23). The system information sent by the BSS on the SACCH contains first information about the neighbor cells and their BCCH (type 5), and second, information about the current cell (Type 6) such as CI and the current LAI.

Channel change

For established RR connections, a channel change within the cell can be performed (dedicated channel assignment; Figure 5.24) to change the configuration of the physical channel in use. Such a channel change can be requested by higher protocol layers, or it can be requested by the RR sublayer; however, it is always initiated by the network. When the MS receives an ASSIGNMENT COMMAND, the transmission of all signaling messages is suspended, the LAPDm connection is taken down, the traffic channel, if it exists, is switched off, and the old channel is deactivated. After activation of the new physical channel and a successful establishment of a new LAPDm connection (Layer 2), the held-back signaling messages can be transmitted.

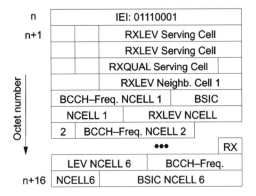

Figure 5.23 Measurement result (IE).

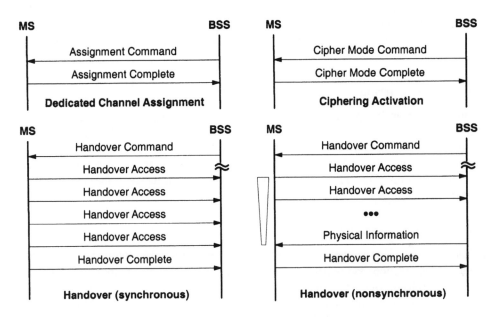

Figure 5.24 Channel change, encryption and handover.

Handover

A second signaling procedure to change the physical channel configuration of an established RR connection is the handover procedure, which is also initiated only from the network side and, for example, becomes necessary if the current cell is left. In contrast to ASSIGNMENT COMMAND, a HANDOVER COMMAND contains not only the new channel configuration but also information about the new cell (e.g. BSIC and BCCH frequency), the procedure variant to establish a physical channel (asynchronous or synchronous handover; Figure 5.24), and a handover reference number.

Having received a HANDOVER COMMAND on the FACCH, the MS terminates the LAPDm connection on the old channel, interrupts the connection, deactivates the old physical channel, and finally switches over to the channel newly assigned in the HANDOVER COMMAND. On the main DCCH (in this case FACCH), the MS sends the unencrypted message HANDOVER ACCESS in an access burst (Figure 4.7, coding as on the RACCH, section 4.8) to the base station. Even though this is a message on the FACCH, an access burst is used because the MS at this time does not yet know the complete synchronization information. The eight data bits of the AB contain the handover reference of the handover command. The way in which these ABs are transmitted depends on whether both cells have synchronized their TDMA transmission or not.

In the case of existing synchronization, the AB (HANDOVER ACCESS) is sent in exactly four subsequent time slots of the main DCCH (FACCH). Thereafter, the MS activates the new physical channel in both directions, establishes an LAPDm connection, activates encryption and, finally, sends a message HANDOVER COMPLETE to the BSS. In the nonsynchronous case, the mobile station repeats the access burst until either a timer expires (handover failed) or until the base station answers with a RR message PHYSICAL INFORMATION which contains the currently needed timing advance and this way enables the establishment of the new RR connection.

Activation of ciphering

Another important RR procedure is the activation of ciphering. This is done by the BSS with the CIPHER MODE COMMAND, which also indicates that the BTS has activated its deciphering function. Having received the CIPHER MODE COMMAND, the MS activates ciphering as well as deciphering and sends the answer CIPHER MODE COMPLETE already in enciphered form. If the BTS is able to correctly decipher this message, the ciphering mode has been successfully established.

Other signaling procedures

In addition, there are a number of less significant signaling procedures defined, such as frequency redefinition, additional assignment, partial release or classmark change. The first concerns the change of the MA; section 4.2.3. The next two deal with a change of the physical channel configuration. With the last message, CLASSMARK CHANGE, the MS reports that it now belongs to a new power class (see Table 4.8), which can be achieved by installing a commercially available power booster kit, for example.

5.4.4 Mobility management

The main task of MM is to support the mobility of the MS; for example, by reporting the current location to the network or verifying the subscriber identity. Another task of the MM sublayer is to offer MM connections and associated services to the CM sublayer above. The message format for MM messages is the uniform Layer 3 signaling message format (Figure 5.21). MM has its own protocol discriminator, and the MM messages are marked with a type code (MT; Table 5.5).

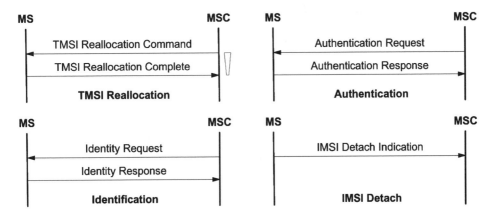

Figure 5.25 MM signaling procedures of category 'common'.

All MM procedures presume an established RR connection, i.e. a dedicated logical channel must be assigned with an established LAPDm connection in place, before MM transactions can be performed. These transactions occur between MS and MSC, i.e. messages are passed through the BSS transparently without interpretation and forwarded to the MSC with the DTAP transport mechanism. The MM procedures are divided into three categories: common, specific, and MM connection management. Whereas common procedures can always be initiated and executed as soon as a RR connection exists, specific procedures exclude one another, i.e. they cannot be processed simultaneously or during a MM connection. Conversely, a MM connection can only be set up if no specific procedure is running.

Common MM procedures

The common MM procedures are summarized in Figure 5.25. In addition to the IMSI detach procedure, they are all initiated from the network side. An important role for the protection of subscriber identity is held by the TMSI reallocation procedure. If the confidentiality of a subscriber's identity IMSI is to be protected (an optional network service), the signaling procedures across the air interface use the TMSI instead of the IMSI. This TMSI has only local significance within a location area and must be used together with the LAI for the unique identification of a subscriber.

For further protection, the TMSI can also be repeatedly reallocated (TMSI reallocation) which must be done at the latest when the location area changes. Otherwise this TMSI change is left as an option to the network operator, but it can be performed any time after an encrypted RR connection to the MS has been set up. The TMSI reallocation is either executed explicitly as a standalone procedure, or implicitly from other procedures using the TMSI, e.g. the location update. In the case of explicit TMSI reallocation, the network sends a TMSI REALLOCATION COMMAND with the new TMSI and the current LAI on an encrypted RR connection to the MS (Figure 5.25).

The MS stores the TMSI and LAI in nonvolatile SIM storage and acknowledges it with the message TMSI REALLOCATION COMPLETE. If this message reaches the MSC before

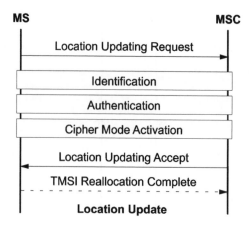

Figure 5.26 MM signaling procedures of category 'specific'.

the timer expires, the timer is cancelled and the TMSI is valid. However, if the timer expires before the acknowledgement arrives, the procedure is repeated. If it fails a second time, the old as well as the new TMSI are barred for a certain time interval and the IMSI is used for paging the MS. If the MS answers a paging call, TMSI reallocation is started again. Furthermore, the TMSI is assumed valid in spite of failed reallocation if it is used by the MS in subsequent transactions.

Two more common procedures are used for the identification of a MS or a subscriber (identification procedure) and for the verification of the respective identity (authentication procedure). For the identification of a MS, there is the equipment identity IMEI as well as the subscriber identity IMSI which is assigned to the MS through the SIM card. The network may request these two identity parameters at any time from the MS with an IDENTITY REQUEST. Therefore the MS must be able at any time to supply these identity parameters to the network with an IDENTITY RESPONSE message.

Authentication also assigns a new key for encryption of user payload data. This procedure is started from the network with an AUTHENTICATION REQUEST message. A mobile station must be able to process this request at any time during a RR connection. The MS calculates the new key Kc for the encryption of user data from the information obtained during authentication which is locally stored, and it also calculates authentication information to prove its identity without doubt. This authentication data are transmitted with an AUTHENTICATION RESPONSE message to the MSC which evaluates them. If the answer is not valid and the authentication has therefore failed, further processing depends on whether the IMSI or TMSI was used. In the case of TMSI, the network can start the identification procedure. If the implied IMSI is not identical to that associated with the TMSI by the network, the authentication is restarted with new correct parameters. If the two IMSIs agree, or the IMSI was used *a priori* by the MS, the authentication has failed, which is indicated to the MS with an AUTHENTICATION REJECT message. This forces the MS to cancel all of the assigned identity and security parameters (TMSI, LAI, Kc) and to enter idle mode, so that only simple cell selection and emergency calls are enabled.

If the MS is powered off or the SIM has been removed, the MS is not reachable because the MS does not monitor the paging channel, and calls cannot be delivered. In order to relieve the paging load on the BSS caused by unnecessary paging calls, a network operator can optionally request an explicit deregistration message from the MS, which is not normally required. This option is signaled by setting a flag on the BCCH (SYSTEM INFORMATION, type 3) and on the SACCH (SYSTEM INFORMATION, type 8). If the flag is set, the MS sends an IMSI DETACH INDICATION message when it powers off or when the SIM is removed, which allows the network to mark the MS as inactive. The IMSI detach procedure is the only common procedure that cannot be started at an arbitrary time even during a specific procedure: its start has to be delayed until any specific procedure has ended.

Specific MM procedures

In GSM systems, updating of current location information is the sole responsibility of the MS. Using the information broadcast on BCCH channels, it has to recognize any change in the current location area and report it to the network, so that the databases HLR and VLR can be kept up to date. The generic structure of a location update is shown in Figure 5.26: the MS requests to update its current location information in the network with a LOCATION UPDATING REQUEST. If this can be done successfully, the network acknowledges this with a message LOCATION UPDATING ACCEPT. In the course of a location update, the network can ask for the MS's identity and check it out (identification and authentication). If the service 'confidential subscriber identity' has been activated, a new TMSI assignment is a permanent component of the location update. In this case, enciphering of user data on the RR connection is activated, and the new TMSI is transmitted together with the message LOCATION UPDATING ACCEPT and is acknowledged with the message REALLOCATION COMPLETE. Periodic updating of location information can be used to indicate the presence of the MS in the network. For this purpose, the MS keeps a timer which periodically triggers a location update procedure. If this option is in use, the timer interval to be used is broadcast on the BCCH (SYSTEM INFORMATION, type 3). The procedure IMSI attach is the converse of the procedure IMSI detach (see Figure 5.25) and is executed as a special variant of the location update if the network requires this. However, the MS executes an IMSI detach only if the LAI broadcast on the BCCH agrees with the LAI stored in the MS. If the stored LAI and received LAI differ, a normal location update procedure is executed.

MM connection management

Finally, there is a third category of MM procedures which are needed for the establishment and the operation of MM connections (Figure 5.27). A MM connection is established on request from the CM sublayer above and serves for the exchange of messages between CM entities, where each CM entity has its own MM connection (Figure 5.12). The procedures for the setup of MM connections are different depending on whether initiation occurs from the network or the MS.

Setting up a MM connection from the side of the MS presumes the existence of a RR connection, but a single RR connection can be used by multiple MM connections. The MM connection can only be established if the MS has executed a successful location update in the current location area. An exception is an emergency call, which is possible at any time. If there is a request from the CM sublayer for a MM connection, it may be delayed or rejected

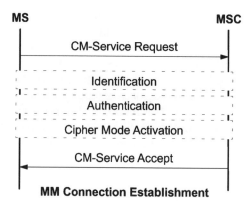

Figure 5.27 MM signaling procedures of category 'MM connection management'.

if there are specific procedures active, depending on implementation. If the MM connection can be established, the MS sends the message CM-SERVICE REQUEST to the network. This message contains information about the mobile subscriber (IMSI or TMSI) as well as information about the requested service (outgoing voice call, SMS transfer, activation or registration of a supplementary service, etc.). Depending on these parameters, the network can execute any common MM procedure (except IMSI detach) or activate enciphering of user data. If the MS receives the message CM-SERVICE ACCEPT or the local message from the RR sublayer that enciphering was activated, it treats this as an acceptance of the service request, and the requesting CM entity is informed about the successful setup of a MM connection. Otherwise, if the service request has been rejected by the network, the MS receives a message CM-SERVICE REJECT and the MM connection cannot be established.

The network-initiated setup of a MM connection does not require an exchange of CM service messages. After successful paging, a RR connection is established, and the sublayer on the network side executes one of the MM procedures if necessary (mostly location update) and requests from the RR sublayer the activation of user data encryption. If these transactions are successful, the service requesting CM entity is informed and the MM connection is established.

5.4.5   Connection management

CC is one of the entities of CM; the CM sublayer is shown in Figure 5.12. It comprises procedures to establish, control and terminate calls. Several parallel CC entities are provided, such that several parallel calls on different MM connections can be processed. For CC, finite state models are defined both on the mobile side as well as on the network side. The two entities at the MS and MSC sites each instantiate a protocol automaton, and these communicate with each other using the messages in Table 5.6 and the uniform Layer 3 signaling message format (Figure 5.21).

Parts of CC in the MS are presented schematically in Figures 5.28 and 5.29. They show the mobile-originating and mobile-terminating setup of a call and the mobile/network initiated call takedown. If there is a desire to call from the MS (mobile-originating call), the CC entity

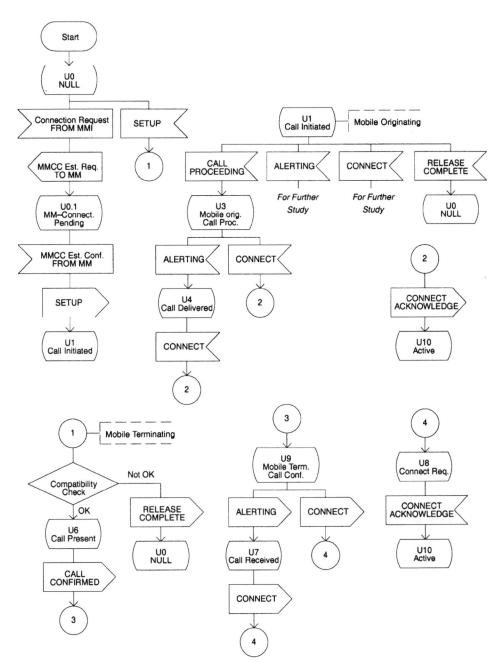

Figure 5.28 Call setup (MS): mobile-originating and mobile-terminating.

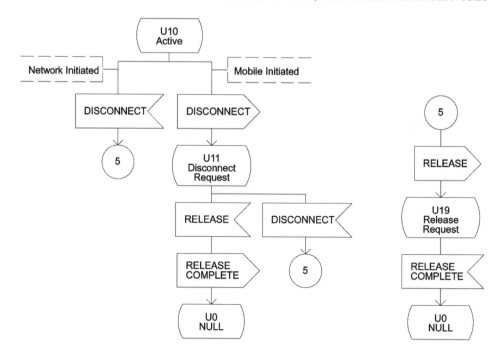

Figure 5.29 Call termination at the MS: mobile-initiated and network-initiated.

first requests a MM connection from the local MM entity, also indicating whether it is a normal or emergency call (MMCC ESTABLISHMENT REQUEST; Figure 5.28). The call to be established on this MM connection requires a special service quality of the MM sublayer.

For a simple call, the MS must be registered with the network, whereas this is only optionally required with an emergency call, i.e. an emergency call can also be established on an unenciphered RR connection from a MS that is not registered.

After successful establishment of this MM connection and activation of the user data encryption, the service-requesting CC entity is informed (further interactions with the MM entity are not shown in Figure 5.28 for call establishment). The MS signals on this connection the desire to connect to the CC entity in the MSC (SETUP). An emergency call is initiated with the message EMERGENCY SETUP; the remaining call setup is identical to that used for single calls.

The MSC can respond to this connection request in several ways: it can indicate with a message CALL PROCEEDING that the call request has been accepted and that all of the necessary information for the setup of the call is available. Otherwise the call request is declined with RELEASE COMPLETE. As soon as the called party is being signaled, the MS receives an ALERTING message; once the called party accepts the call, a CONNECT message is returned which is acknowledged with a CONNECT ACKNOWLEDGE message, thus switching through the call and the associated user data connection. If the call request need not be signaled to the called party and the call can be accepted directly, the ALERT message is omitted. CC signaling in GSM corresponds essentially to the call setup according to Q.931 in ISDN. In addition, CC in GSM has a number of peculiarities, especially to account for the limited resources and properties of the radio channel. In particular, the call request of the MS can be entered into a queue (call queuing), if there is no immediately free traffic channel (TCH) for the establishment of the call. The maximum waiting time a call may have to wait for assignment of a traffic channel can be adjusted according to operator needs. Furthermore, the point at which the traffic channel is actually assigned can be chosen. For example, the traffic channel can be assigned immediately after acknowledging the call request (CALL PROCEEDING); this is early assignment. On the other hand, the call can be first processed completely and the assignment occurs only after the targeted subscriber is being called; this is late assignment or Off-Air Call Setup (OACSU). The variant OACSU avoids unnecessary allocation of a traffic channel if the called subscriber is not available. The blocking probability for call arrivals at the air interface can be reduced this way. On the other hand, there is the probability that after a successful call request signaling procedure, no traffic channel can be allocated for the calling party before the called party accepts the call, and thus the call cannot be completely switched through and must be broken off.

If a call arrives at the MS (mobile-terminating call), a RR connection with the MS is established within the MM connection setup (inclusive of paging). Once the MM connection is successfully completed and the encryption is activated, the call request is signaled to the MS with a SETUP message. This message includes information about the requested service, and the MS examines first whether it can satisfy the requested service profile (compatibility check). If affirmative, it accepts the call request and signals this to the local subscriber (local generation of call signal). This is finally communicated to the MSC with a CALL CONFIRMED message and an ALERTING message. If the mobile subscriber eventually accepts the call, the call is switched through completely with handshake messages CONNECT and CONNECT ACKNOWLEDGE. If, because of the selected service, there is no necessity for call request signaling to the called subscriber and the call can be switched through immediately (e.g. with a fax call), the MS signals the call acceptance (CONNECT) immediately after the message CALL CONFIRMED. Call queuing and OACSU can also be used for mobile-terminating calls. The OACSU variant for mobile-terminating calls allocates a traffic channel only after the call has been accepted by the mobile subscriber with a CONNECT message.

The release of a connection is started with a DISCONNECT message either from the mobile or the network side (mobile-/network-initiated) and is completed with handshake messages RELEASE and RELEASE COMPLETE. If there is a collision of DISCONNECT messages, i.e. if both CC entities send a DISCONNECT simultaneously, they also answer it with a RELEASE so that a secure termination is ensured.

During an established call, two more CC procedures can be employed: DTMF signaling and in-call modification. DTMF signaling is an inband signaling procedure, which allows terminals (here MSs) to communicate with the respective other side, e.g. answering machines, or configuring special network services, e.g. voice mailboxes in the network. In GSM, DTMF can only be used during a voice connection. With a message START DTMF on the FACCH, the network is told that a key has been pressed, and the release of the key is signaled with a STOP DTMF message (Figure 5.30). Each of these messages is acknowledged by the network (MSC). A minimum interval must be maintained between the START/STOP messages ($T_{\text{press\_min}}$, $T_{\text{release\_min}}$). While a key is depressed at the MS, the MSC generates a DTMF tone corresponding to the key code signaled with the START DTMF

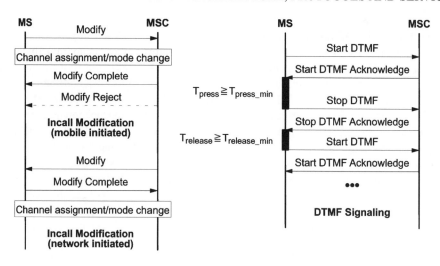

Figure 5.30 DTMF signaling and service change.

command. The DTMF tones must be generated within the MSC, since the speech coding in the GSM codec does not permit the pure transmission of DTMF tones in the voice band, and thus DTMF tones generated by the MS would arrive at the other side in distorted form.

Using the in-call modification procedure, a service change can be performed, e.g. when speech and fax data are sent in sequence and are alternating during a call. A service change is started either from the MS or from the network by sending a MODIFY request. This request contains the service and kind of change (returning or nonreturning). After sending the MODIFY request, the transmission of user data is halted. If the service change can be performed, this is signaled with a MODIFY COMPLETE message; otherwise the request is denied with a MODIFY REJECT. The service change may necessitate a change in the current physical channel configuration or operating mode. For this purpose, the MSC will use the respective channel assignment (ASSIGNMENT COMMAND; Figure 5.24).

5.4.6   Structured signaling procedures

The preceding sections have presented the basic signaling procedures of the three sublayers RR, MM and CM. These procedures have to cooperate in the form of structured procedures for the different transactions. The elements of a structured signaling procedure are as follows.

- Phase 1: paging, channel request, assignment of a signaling channel.

- Phase 2: service request and collision resolution.

- Phase 3: authentication.

- Phase 4: activation of user data encryption.

- Phase 5: transaction phase.

- Phase 6: release and deallocation of the channel.

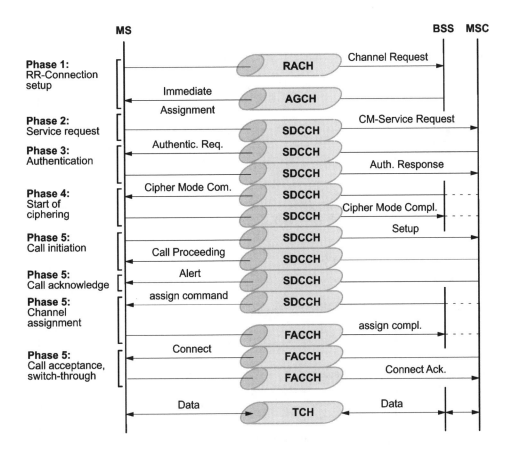

Figure 5.31 Mobile-initiated call setup with OACSU (late assignment).

Two examples of structured signaling procedures are presented in Figures 5.31 and 5.32. They show the phases executed for the structured transaction, the terminating entities (MS, BSS, MSC), the respective message and the logical channel used for the transport of the message. The first example (Figure 5.31) is a mobile-initiated call setup with OACSU – a traffic channel is only assigned after the subscriber of the called station is presented with the call request (ALERTING). The second example (Figure 5.32) shows a service change from voice to data and the modification of the selected data service. Such a modification could, for example, be the change of transmission rate. Finally, the call is released and the traffic channel is deallocated.

5.4.7 Signaling procedures for supplementary services

As can be seen in Figure 5.21, signaling messages to control SS are coded with special protocol discriminators: 0011 for call related; 1011 for noncall related. A special set of signaling messages has been defined for their control (Table 5.7). The category CC

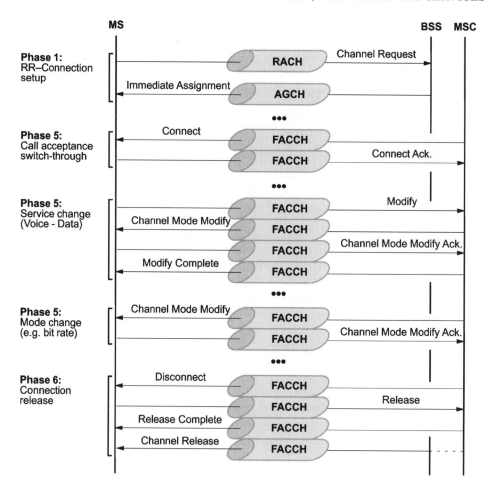

Figure 5.32 In-call modification and call release.

Messages (Table 5.6) consists of the subcategories call information phase (message type MT = 0x01tttt) and Miscellaneous (message type MT = 0x11tttt). These two message categories are used in two categories of SS procedures: the separate message approach and the common information element procedure. Whereas the separate message approach uses its own messages (HOLD/RETRIEVE; Table 5.7) to activate specific functions, the functions of the common information element procedure are handled with a generic FACILITY message. Functions of the first category need synchronization between network and MS. The FACILITY category, however, is only used for SS which do not require synchronization. This distinction becomes obvious in the examples of realized SS, which is presented in the following.

The messages of the separate message approach can be used during the call information phase to realize supplementary services like hold, callback or call waiting. Figure 5.33 shows

Table 5.7 CC messages for SS.

Category	Message	Direction	MT
Call information phase	Hold	N ↔ MS	0x011000
	Hold acknowledge	N ↔ MS	0x011001
	Hold reject	N ↔ MS	0x011010
	Retrieve	N ↔ MS	0x011100
	Retrieve acknowledge	N ↔ MS	0x011101
	Retrieve reject	N ↔ MS	0x011110
Miscellaneous	Facility	N ↔ MS	0x111010
	Register	N ↔ MS	0x111011

examples. A completely established call (call reference CR: 1 in Figure 5.33) can be put into the hold state from either one of the two partner entities.

To perform this SS, it is initiated with a HOLD message. The MSC interrupts the connection and indicates with a HOLD message to the partner entity that the call is in the hold state. On each call segment this fact is acknowledged with a HOLD ACKNOWLEDGE message, which leads both the requesting MS and the MSC to cut the traffic channel. The MS which caused the hold state to be entered can now establish another call (CR: 2 in Figure 5.33) or accept a call that may be coming in. Using another handshake HOLD/HOLD ACKNOWLEDGE, this call could also be put into the hold state and there could be switching between both held calls (brokering). For this purpose, a held call (CR: 1 in Figure 5.33) can be reactivated with a RETRIEVE message and reconnected to the call at each side of the traffic channel after the reactivation of the call has been acknowledged with a RETRIEVE ACKNOWLEDGE message.

These call-related signaling procedures modify the call state and define an extended state diagram with an auxiliary state. The participating calls all remain in the active state whereas the auxiliary state changes between hold and idle. In a two-dimensional state space, for example, call CR: 1 changes from (active, idle) through the state (active, hold request) into the state (active, call held) and back through the state (active, retrieve request).

If outgoing or incoming calls are barred (Figure 5.34), a call request is immediately refused by giving a RELEASE COMPLETE message with a reason in a FACILITY information element (Barring Of Outgoing Calls (BAOC), Barring Of Incoming Calls (BAIC)). A state change of the call in the extended state space does not occur. The assumption is, of course, that the calling or called subscriber has activated call barring. The MSC receiving the call request from the calling subscriber must verify the activation of this supplementary service. This requires an inquiry of the HLR of the calling subscriber (for BAOC) or the called subscriber (for BAIC), since the HLRs store the service profiles of the respective subscribers. In this case the HLR acts not only as a database but also as a participant in controlling intelligent network services.

Another call-related supplementary service uses the FACILITY message of the common information element procedure: Call Forwarding Unconditional (CFU); see Figure 5.34. With this supplementary service, a regular call setup is performed, however, not to the

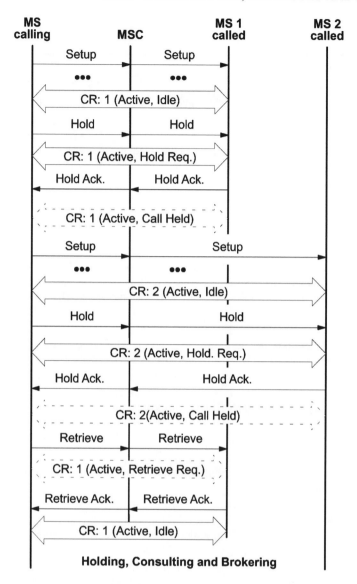

Figure 5.33 Call holding and associated procedures.

called subscriber but to the forwarding target selected when the service was activated (in Figure 5.34 it is another MS). The calling subscriber is informed about the change of the called number with a FACILITY message. Likewise the target of the forwarding is informed with a FACILITY message that the incoming call is a forwarded call. In this case, there is no necessity for a change in the extended state diagram nor is synchronization between network and MS required. It is only necessary that the involved MSs are informed about the occurrence of forwarding. In this case, the target of the call forwarding is also stored in

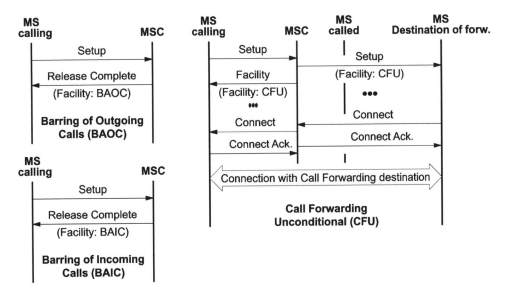

Figure 5.34 Barring and forwarding of calls.

the HLR of the subscriber who activated the service (the called MS in Figure 5.34). Thus, the call processing in the MSC of the called subscriber must be interrupted and the HLR must be informed about the call request. If the called subscriber has activated unconditional call forwarding, the HLR returns the new call target to the MSC, which can continue call processing with the changed target number.

5.4.8 Realization of SMS

The procedures for the transport of point-to-point short messages reside in the CM sublayer, also called the Short Message Control Layer (SM-CL) and in the Short Message Relay Layer (SM-RL) directly above. Accordingly, the protocol entities are called the Short Message Control (SMC) entity and the Short Message Relay (SMR) entity. A complete established MM connection is needed for the transport of short messages, which again presumes an existing RR connection with LAPDm protection on a SDCCH or SACCH channel. To distinguish among these packet-switched user data connections, SMS messages are transported across SAPI = 3 of the LAPDm entity.

A SMS transport PDU (SMS-SUBMIT or SMS-DELIVER; Figure 5.35) is transmitted with an RP-DATA message between MSC and MS using the SM-RP; section 5.3.2. Correct reception is acknowledged with an RP-ACK message either from the SMS-SC (mobile-initiated SMS transfer) or from the MS (mobile-terminated SMS transfer).

For the transfer of short messages between SMR entities in MS and MSC, the CM sublayer provides a service to the SM-RL layer above. The SMR entity requests this service for the transfer of RP-DATA or RP-ACK (MNSMS-ESTABLISH-REQUEST; Figure 5.36). Following the SMR service request, the SMC entity itself requests a MM connection on which it then transfers the short message inside a CP-DATA message. The appropriate

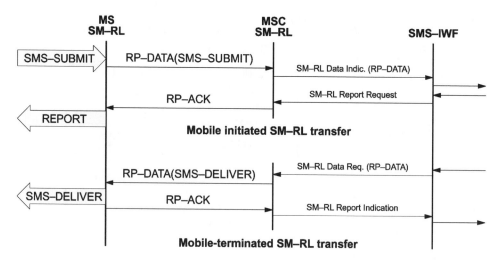

Figure 5.35 Short message transfer between SMR entities.

service primitives between protocol layers are also illustrated in Figure 5.36. The correct reception of CP-DATA is acknowledged with CP-ACK. In these SMC messages, one PDU transports a SDU from the SMR sublayer above. This SMC-SDU is the SMS relay message RP-DATA and its acknowledgement RP-ACK which are used to signal the transfer of short messages. The Short Message Transport Layer (SM-TL) above the SM-RL provides end-to-end transport of short messages between the MS and SMS-SC.

5.5   Signaling at the A and Abis interfaces

Whereas the transport of user data between MSC and BSC occurs across standard connections of the fixed network with 64 or 2048 kbit/s (or 1544 kbit/s), the transport of signaling messages between MSC and BSC runs over the SS#7 network. The MTP and SCCP parts of SS#7 are used for this purpose. A protocol function using the services of the SCCP is defined at the A interface. This is the BSSAP, which is further subdivided into DTAP and BSSMAP; see Figure 5.37. In addition, the Base Station System Operation and Maintenance Part (BSSOMAP) was introduced, which is needed for the transport of network management information from OMC via the MSC to the BSC.

At the A interface, one can distinguish between two signaling message streams: one between MSC and MS and another between MSC and BSS. The messages to the MS (CM, MM) are passed on transparently through the BSS using the DTAP protocol part of SS#7. BSC and BTS do not interpret them. The SCCP protocol part provides a connection-oriented and a connectionless transfer service for signaling messages. For DTAP messages, only a connection-oriented service is offered. The DTAP of the BSSAP uses one signaling connection for each active MS with one or more transactions per connection. A new connection is established each time when messages of a new transaction with a MS are to be transported between MSC and BSS.

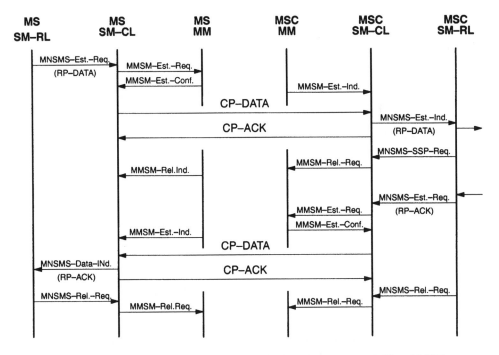

Figure 5.36 Short message transfer on the CM plane between MS and MSC.

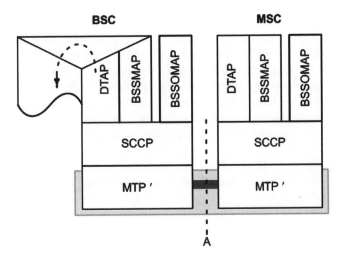

Figure 5.37 Protocols at the A interface between MSC and BSS.

Two cases of setting up a new SCCP connection are distinguished. First, in the case of location update and connection setup (outgoing or incoming), the BSS requests an SCCP

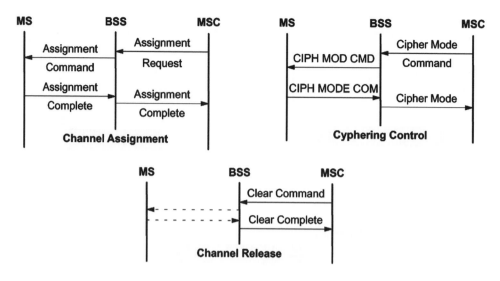

Figure 5.38 Examples of dedicated BSSMAP procedures.

connection after the channel request (access burst on RACH) from the MS has been satisfied with a SDCCH or TCH and after an LAPDm connection has been set up on the SDCCH or FACCH. The second situation for setting up an SCCP connection is a handover to an other BSS, in which case the MSC initiates the connection setup. Most of the signaling messages at the air interface (CM and MM; Tables 5.5 and 5.6) are passed transparently through the BSS and packaged into DTAP-PDUs at the A interface, with the exception of some RR messages.

The BSSMAP implements two more kinds of signaling procedure between MSC and BSS: first those concerning one MS or single physical channels at the air interface; and, second, global procedures for the control of all of the resources of a BSS or cell. In the first case, the BSSMAP also uses connection-oriented SCCP services, whereas, in the second case, global procedures are performed with connectionless SCCP services. Among the BSSMAP procedures for a dedicated resource of the air interface are functions of resource management (channel assignment and release, start of ciphering) and of handover control (Figures 5.38 and 5.39).

Among the global procedures of BSSMAP are paging, flow control to prevent overloading protocol processors or CCCH channels, closing and opening of channels, and parts of handover control (Figure 5.40).

The transmission layer at the Abis interface between BTS and BSC is usually realized as a primary multiplexed line with 2048 kbit/s (1544 kbit/s in North America) or 64 kbit/s. This may include one physical connection per BTS or one for each connection between a TRX/BCF module and the BTS (Figure 5.5). On these digital paths, traffic or signaling channels of 16 or 64 kbit/s are established. The Layer 2 protocol at the Abis interface is LAPDm, whose Terminal Equipment Identifier (TEI) is used to address the TRX and/or BCF of a BTS (Figure 5.41).

Several LAPDm connections are established for each TEI: the Radio Signaling Link (RSL), SAPI = 0; the Operation and Maintenance Link (OML), SAPI = 62; and the Layer 2

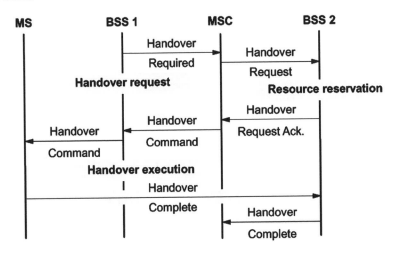

Figure 5.39 Dedicated BSSMAP procedures for internal handover.

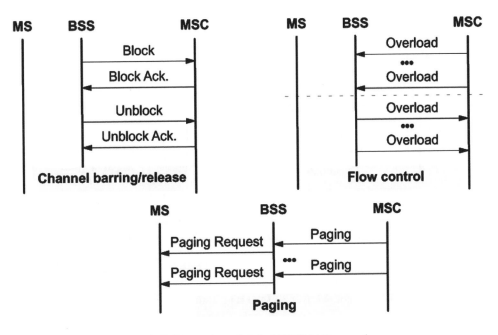

Figure 5.40 Examples of global BSSMAP procedures.

Management Link (L2ML), SAPI = 63. Traffic management is handled on the RSL, operation and maintenance on the OML, and management messages of Layer 2 are sent on the L2ML to the TRX or BCF. The RSL is the most important of these three links for the control of radio resources and connections for communication between MS and

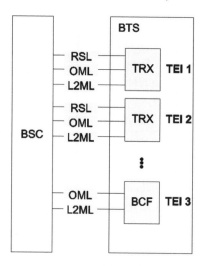

Figure 5.41 Logical connections at Layer 2 of the Abis interface.

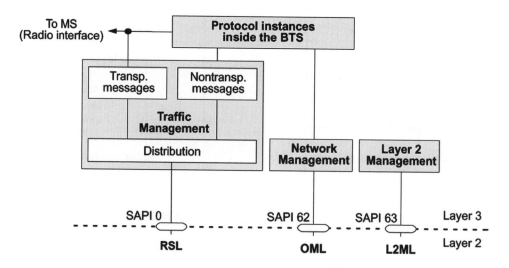

Figure 5.42 Protocol Layer 3 of the BTS at the Abis interface (BTSM).

network. Two types of messages are distinguished on this signaling link: transparent and nontransparent messages (Figure 5.42). Whereas the BTS passes on transparent messages from/to the LAPDm entity of a MS without interpreting or changing them, nontransparent messages are exchanged between BTS and BSC.

In addition, one distinguishes between four groups of traffic management messages of the BTS.

Figure 5.43 Transfer of transparent signaling messages.

- *Radio link layer management*: procedures to establish, modify and release connections of the link layer (LAPDm) to the MS at the air interface Um.

- *Dedicated channel management*: procedures to start ciphering, transfer of channel measurement reports of a MS, transmitter power control of MS and BTS, handover detection and modification of a dedicated channel of the BTS for a specific MS which can then receive the channel in another message (assign, handover command).

- *Common channel management*: procedures for transferring channel requests from MS (received on RACH), start of paging calls, measurement and transfer of CCCH traffic load measurements, modification of BCCH broadcast information, channel assignments to the MS (on the AGCH) and transmission of SMSCBs.

- *TRX management*: procedures for the transfer of measurements of free traffic channels of a TRX to the BSC or for flow control. In the case of overloaded TRX processors or overload on the downlink CCCH/ACCH.

In this way, all of the RR functions in the BTS can be controlled. The majority of RR messages (Table 5.4) are passed on transparently and do not terminate in the BTS. These messages are transported between BTS and BSC (Figure 5.43) in special messages (DATA REQUEST/INDICATION) packaged into LAPDm frames (Layer 2 at the radio interface).

All protocol messages received by the BTS on the uplink from the MS in LAPDm I/UI frames, except for channel measurement reports of the MS, are passed on as transparent messages in a DATA INDICATION.

Except for the link protocol LAPDm which is completely implemented in the BTS, there are some functions which are also handled by the BTS, and the pertinent messages from or to the MS are transformed by the BTS into the appropriate RR messages. This includes channel assignment, ciphering, assembly of channel measurements from MS and TRX and their transfer to the BSC (possibly with processing in the BTS), power control commands from the BTS for the MS, and channel requests from the MS (on the RACH) as well as channel assignments (Figure 5.44). Thus, four of the RR messages on the downlink direction to the MS (Table 5.4) cannot be treated as transparent messages: CIPHERING MODE COMMAND, PAGING REQUEST, SYSTEM INFORMATION and the three IMMEDIATE ASSIGN messages. All of the other RR messages to the MS are sent transparently within a DATA REQUEST to the BTS.

Figure 5.45 shows the format of a BTSM message (Layer 3 between BSC and BTS). Transparent and nontransparent messages are distinguished between with a message discriminator in the first octet. For this purpose, the T-bit (bit 1 of octet 1) is set to logical 1 for messages which the BTS is supposed to handle transparently or has recognized as transparent.

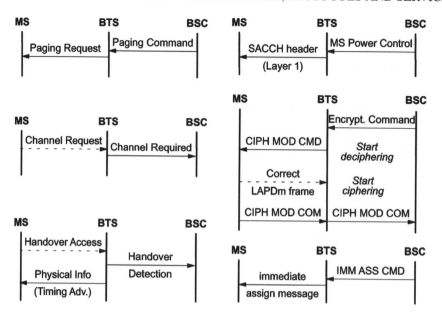

Figure 5.44 Examples of nontransparent signaling between BTS and BSC.

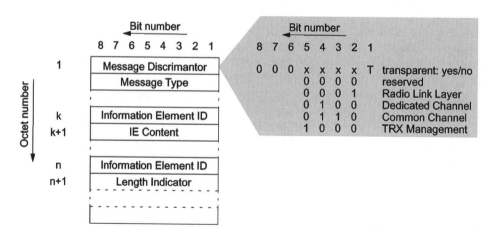

Figure 5.45 Format of BTSM-RSL protocol messages.

Bits 2 to 5 serve to assign the messages to one of the four groups defined on the RSL. Including the MT defines the message completely (Figure 5.45). The remainder of the BTSM message contains mandatory and optional IEs which have a fixed length of mostly two octets or which contain an additional length indicator in the case of variable length.

5.6   Security-related network functions: authentication and encryption

Methods of encryption for user data and for the authentication of subscribers, like all techniques for data security and data protection, are gaining enormous importance in modern digital systems (Eberspächer, 1999). GSM therefore introduced powerful algorithms and encryption techniques. The various services and functions concerned with security in a GSM PLMN are categorized in the following way:

- subscriber identity confidentiality;

- subscriber identity authentication;

- signaling information element confidentiality;

- data confidentiality for physical connections.

In the following, the security functions concerning the subscriber are presented.

5.6.1   Protection of subscriber identity

This function is intended to prevent the disclosure of which subscriber is using which resources in the network, by listening to the signaling traffic on the radio channel. On the one hand, this should ensure the confidentiality of user data and signaling traffic; on the other hand, it should also prevent the localization and tracking of a MS. This means above all that the IMSI should not be transmitted as clear text, i.e. unencrypted.

Instead of the IMSI, one uses a TMSI on the radio channel for identification of subscribers. The TMSI is temporary and has only local validity, which means that a subscriber can only be uniquely identified by TMSI and the LAI. The association between IMSI and TMSI is stored in the VLR.

The TMSI is issued by the VLR, at the latest, when the MS changes from one LA into another (location updating). When a new location area is entered, this is noticed by the MS (section 3.3.5) which reports to the new VLR with the old LAI and TMSI (LAI*old* and TMSI*old*; Figure 5.46). The VLR then issues a new TMSI for the MS. This TMSI is transmitted in encrypted form.

The subscriber identity is thus protected against eavesdropping in two ways: first, the TMSI is used on the radio channel instead of the IMSI; second, each new TMSI is transmitted in encrypted form.

In the case of database failures, if the VLR database is partially lost or no correct subscriber data are available (loss of TMSI, TMSI unknown at VLR, etc.), the GSM standard provides for a positive acknowledgement of the subscriber identity. For this subscriber identification, the IMSI must be transmitted as clear text (Figure 5.47) before encryption is turned on. Once the IMSI is known, encryption can be restarted and a new TMSI can be assigned.

5.6.2   Verification of subscriber identity

When a subscriber is added to a home network for the first time, a subscriber authentication key (Ki) is assigned in addition to the IMSI to enable the verification of the subscriber identity

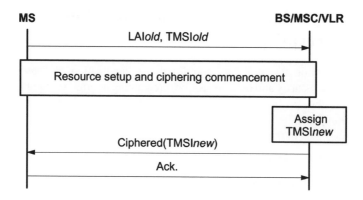

Figure 5.46 Encrypted transmission of the temporary subscriber identity.

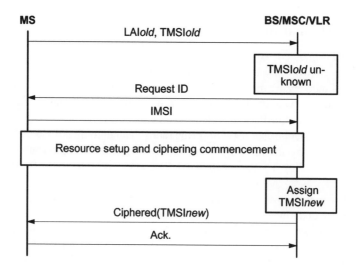

Figure 5.47 Clear text transmission of the IMSI when the TMSI is unknown.

(also known as authentication). All security functions are based on the secrecy of this key. At the network side, the key Ki is stored in the Authentication Center (AUC) of the home PLMN. At the subscriber side, it is stored on the SIM card of the subscriber.

The process of authenticating a subscriber is essentially based on the A3 algorithm, which is performed at the network side as well as at the subscriber side (Figure 5.48). This algorithm calculates independently on both sides (MS and network) the Signature Response (SRES) from the authentication key Ki and a Random Number (RAND) offered by the network. The MS transmits its SRES value to the network which compares it with its calculated value. If both values agree, the authentication was successful. Each execution of the algorithm A3 is performed with a new value of the random number RAND which cannot be predetermined;

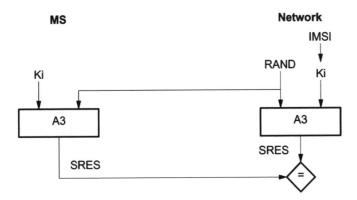

Figure 5.48 Principle of subscriber authentication.

in this way recording the channel transmission and playing it back cannot be used to fake an identity.

5.6.3   Generating security data

At the network side, the 2-tuple (RAND, SRES) need not be calculated each time when authentication has to be done. Rather the AUC can calculate a set of (RAND, SRES) 2-tuples in advance, store them in the HLR, and send them on demand to the requesting VLR. The VLR stores this set (RAND[n], SRES[n]) and uses a new 2-tuple from this set for each authentication procedure. Each 2-tuple is used only once; so new 2-tuples continue to be requested from the HLR/AUC.

This procedure, which lets the security data (Kc, RAND, SRES) be calculated in advance by the AUC, has the advantage that the secret authentication key Ki of a subscriber can be kept exclusively within the AUC, which ensures a higher level of confidentiality. A somewhat less secure variant is to supply the currently needed key Ki to the local VLR which then generates the security data locally.

If the key Ki is kept exclusively in the AUC, the AUC has to generate a set of security data for a specific IMSI on demand from the HLR (Figure 5.49): the random number RAND is generated and the pertinent signature SRES is calculated with the A3 algorithm, whereas the A8 algorithm generates the encryption key Kc.

The set of security data, a 3-tuple consisting of Kc, RAND, and SRES, is sent to the HLR and stored there. In most cases, the HLR keeps a supply of security data (e.g. 5), which can then be transmitted to the local VLR, so that one does not have to wait for the AUC to generate and transmit a new key. When there is a change of LA into one belonging to a new VLR, the sets of security data can be passed on to the new VLR. This ensures that the subscriber identity IMSI is transmitted only once through the air, namely when no TMSI has yet been assigned (see registration) or when these data have been lost. Afterwards the (encrypted) TMSI can be used for communicating with the MS.

If the IMSI is stored on the network side only in the AUC, all authentication procedures can be performed with the 2-tuples (RAND, SRES) which were precalculated by the AUC.

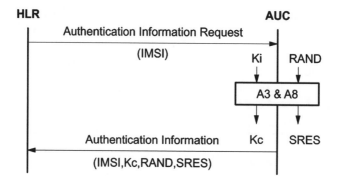

Figure 5.49 Generation of a set of security data for the HLR.

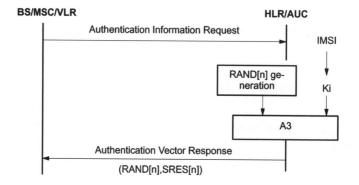

Figure 5.50 Highly secure authentication (no transmission of Ki).

In addition to relieving the load on the VLR (no execution of the A3 algorithm), this kind of subscriber identification (Figure 5.50) has the other advantage of being particularly secure, because confidential data, especially Ki, need not be transmitted over the air. It should be used, in particular, when the subscriber is roaming in a network of a foreign operator, since it avoids passing of security-critical data over the network boundary.

The less secure variant (Figure 5.51) should only be used within a PLMN. In this case, the secret (security-critical) key Ki is transmitted each time from the HLR/AUC to the current VLR, which executes the algorithm A3 for each authentication.

5.6.4 Encryption of signaling and payload data

The encryption of transmitted data is a special characteristic of GSM networks that distinguishes the offered service from analog cellular and fixed ISDN networks. This encryption is performed at the transmitting side after channel coding and interleaving and immediately preceding modulation (Figure 5.52). On the receiving side, decryption directly follows the demodulation of the data stream.

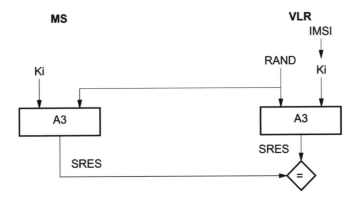

Figure 5.51 Weakly secure authentication (transmission of Ki to VLR).

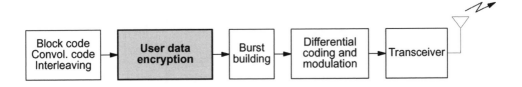

Figure 5.52 Encryption of payload data in the GSM transport chain.

A Cipher Key (Kc) for the encryption of user data is generated at each side using the generator algorithm A8 and the random number RAND of the authentication process (Figure 5.53). This key Kc is then used in the encryption algorithm A5 for the symmetric encryption of user data. At the network side, the values of Kc are calculated in the AUC/HLR simultaneously with the values for SRES. The keys Kc are combined with the 2-tuples (RAND, SRES) to produce 3-tuples, which are stored at the HLR/AUC and supplied on demand, in case the subscriber identification key Ki is only known to the HLR (section 5.6.2). In the case of the VLR having access to the key Ki, the VLR can calculate Kc directly.

The encryption of signaling and user data is performed at the MS as well as at the base station (Figure 5.54). This is a case of symmetric encryption, i.e. ciphering and deciphering are performed with the same key Kc and the A5 algorithm.

Based on the secret key Ki stored in the network, the cipher key Kc for a connection or signaling transaction can be generated at both sides, and the BTS and MS can decipher each other's data. Signaling and user data are encrypted together (TCH/SACCH/FACCH); for dedicated signaling channels (SDCCH) the same method is used as for traffic channels. This process is also called a *stream cipher*, i.e. ciphering uses a bit stream which is added bitwise to the data to be enciphered (Figure 5.55).

Deciphering consists of performing an additional EXCLUSIVE OR operation of the enciphered data stream with the ciphering stream. The FN of the current TDMA frame within a hyperframe (section 4.3.1) is another input for the A5 algorithm besides the key

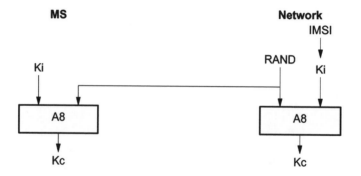

Figure 5.53 Generation of the cipher key Kc.

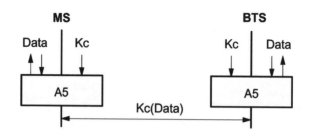

Figure 5.54 Principle of symmetric encryption of user data.

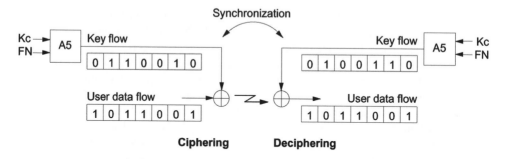

Figure 5.55 Combining payload data stream and ciphering stream.

Kc, which is generated anew for each connection or transaction. The current frame number is broadcast on the SCH and is thus available any time to all mobile stations currently in the cell. Synchronization between ciphering and deciphering processes is thus performed through FN.

However, the problem of synchronizing the activation of the ciphering mode has to be solved first: the deciphering mechanism on one side has to be started at precisely the correct moment. This process is started under network control, immediately after the authentication procedure is complete or when the key Kc has been supplied to the base station; Figure 5.56.

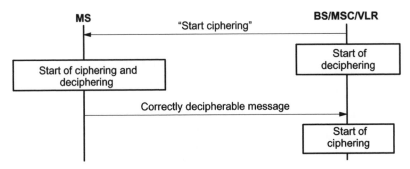

Figure 5.56 Synchronized start of the ciphering process.

The network, i.e. the BTS, transmits to the MS the request to start its (de)ciphering process, and it starts its own deciphering process. The MS then starts its ciphering and deciphering. The first ciphered message from the MS which reaches the network and is correctly deciphered leads to the start of the ciphering process on the network side.

5.7   Signaling at the user interface

Another often neglected but nevertheless very important interface in a mobile system is the user interface of the MS equipment. This Man–Machine Interface (MMI) can be realized freely and therefore in many different ways by the mobile equipment manufacturers. In order to keep a set of standardized service control functions in spite of this variety, the MMI commands have been introduced. These MMI commands define procedures mainly for the control of basic services and SS. The control procedures are constructed around the input of command token strings which are delineated and formatted with the tokens * and #. In order to avoid a user having to learn and memorize a certain number of service control procedures before being able to use the mobile phone, a small set of basic required commands for the MMI interface has been defined; this is the basic public MMI which must be satisfied by all MSs.

The specification basic public MMI outlines the basic functions which must be implemented as a minimum at the MMI of a MS. This includes the arrangement of a 12-key keyboard with the numbers 0 through 9 and the keys for * and # as well as SEND and END keys, which also serve to initiate a desired call or accept or terminate a call.

Some basic operational sequences for making or taking a call are also defined. These requirements are so general that they can be easily satisfied by all mobile equipment.

The MMI commands for the control of SS and the enquiry and configuration of parameters are much more extensive. Using a set of MMI commands which are uniform for all MSs allows control functions to be performed which are often hidden in equipment-specific user guidance menus. For certain functional areas, a MS can thus be operated in a manufacturer-independent way, if one forgoes the sometimes very comfortable possibilities of user-guiding menus and instead learns the control sequences for the respective functions. These sequences are mapped onto the respective signaling procedures within the MS.

Table 5.8 Input format of some MMI commands.

Function	MMI procedure
Activate	*nn(n)*Si#
Deactivate	#nn(n)*Si#
Status enquiry	*#nn(n)*Si#
Registration	**nn(n)*Si#
Delete	##nn(n)*Si#

A MMI command is always constructed according to the same pattern. Five basic formats are distinguished (Table 5.8), which all start with a combination of the tokens * and #: activation (*), deactivation (#), status inquiry (*#), registration (**) and cancellation (##).

In addition, the MMI command must contain a MMI service code of two or three tokens, which selects the function to be performed. In certain cases, the MMI procedure requires additional arguments or parameters, which are separated by *, as Supplementary Information (Si). The MMI command is always terminated with # and may also require the SEND key to be depressed, if the command is not executed locally within the MS but must be transmitted to the network. Table 5.9 contains some basic examples of MMI commands, e.g. the inquiry for the IMEI of the MS (*# 06#) or the change of the PIN (**04*old_PIN*new_PIN*new_PIN#) used to protect the SIM card against misuse. This example also shows how supplementary information is embedded in the command.

Table 5.9 Some basic MMI commands.

Function	MMI procedure
Mobile phone IMEI enquiry	*#06#
Change password for call barring	**03*330*old_PWD*new_PWD*new_PWD#
Change PIN in SIM	**04*old_PIN*new_PIN*new_PIN#
Select SIM number storage	n(n)(n)#

With MMI commands, it is also possible to configure and use SS. For this purpose, each SS is designated with a two-or three-digit MMI service code to select the respective SS (Table 5.10). In some cases, supplementary information is mandatory for the activation of the service, e.g. one needs the target Destination Number (DN) for the call forwarding functions, or the Activation Password (PW) for the SS of barring incoming or outgoing calls (Sia in Table 5.10).

The example of unconditional call forwarding also illustrates the difference between registration and activation of a service. With the command **21*call_number# the forwarding function is registered, the target number configured, and the unconditional forwarding activated. Later, the unconditional forwarding can be deactivated any time with #21# and

Table 5.10 MMI service codes for SS.

Abbreviation	Service	MMI code	Sia	Sib
	All call forwarding, only for (de)activation	2	–	–
	All conditional call forwarding (not CFU), only for (de)activation	4	–	–
CFU	Call forwarding unconditional	21	DN	BS
CFB	Call forwarding on mobile subscriber busy	67	DN	BS
CFNRy	Call forwarding on no reply	61	DN	BS
CFNRc	Call forwarding on mobile subscriber not reachable	62	DN	BS
	All call barring (only for deactivation)	330	PW	BS
BAOC	Barring of all outgoing calls	33	PW	BS
BOI	Barring of outgoing international calls	331	PW	BS
BOIC-exHC	Barring of outgoing international calls except those to home PLMN	332	PW	BS
BAIC	Barring of all incoming calls	35	PW	BS
BIC-Roam	Barring of incoming calls when roaming outside the home PLMN	351	PW	BS
CLIP	Calling line identification presentation	30	–	BS
CLIR	Calling line identification restriction	31	–	BS
CW	Call waiting	43	–	BS
COLP	Connected line identification presentation	76	–	BS
COLR	Connected line identification restriction	77	–	BS

reactivated with *21#. The target number call_number remains stored, unless it is cancelled with the command ##21#. After cancellation, if call forwarding is desired again, it must first be registered again using the **21... command. For some basic services, characteristics can also be activated selectively. The MMI command can contain a second parameter field

Table 5.11 MMI codes for basic services.

Category	Service	MMI code BS
Telematic service	All telematic services	10
	Telephone	11
	All data services	12
	Facsimile	13
	Videotex	14
	Teletext	15
	SMS	16
	All data services except SMS	18
	All telematic services except SMS	19
Bearer service	All bearer services	20
	All asynchronous services	21
	All synchronous services	22
	All connection-oriented synchronous data services	24
	All connection-oriented asynchronous data services	25
	All packet-oriented synchronous data services	26
	All PAD-access services	27

with supplementary information (Sib, Table 5.10) which is again delineated with *. This field contains the service code BS of the basic service for which the SS is to become effective.

An overview of MMI codes for basic services is given in Table 5.11. For example, one can bar incoming calls except short messages with the command **35*PW*18#, or one can forward incoming fax calls to the number fax_number with the command **21*fax_number*13# (the other teleservices remain unaffected).

6

Roaming and handover

6.1 Mobile application part interfaces

The main benefit for the mobile subscribers that the international standardization of GSM has brought is that they can move freely not only within their home networks but also in international GSM networks and that at the same time they can even gain access to the special services they subscribed to at home provided that there are agreements between the operators. The functions needed for this free roaming are called roaming or mobility functions. They rely mostly on the GSM-specific extension of the SS#7. The MAP procedures relevant for roaming are first the location registration/update, IMSI attach/detach, requesting subscriber data for call setup and paging. In addition, the MAP contains functions and procedures for the control of SS and handover, for subscriber management, for IMEI management, for authentication and identification management, as well as for the user data transport of the SMS. MAP entities for roaming services reside in the MSC, HLR and VLR. The corresponding MAP interfaces are defined as B (MSC-VLR), C (MSC-HLR), D (HLR-VLR), E (MSC-MSC) and G (VLR-VLR) (Figure 3.6). At the subscriber interface, the MAP functions correspond to the functions of MM, i.e. the MM messages and procedures of the Um interface are translated into the MAP protocols in the MSC.

The most important functions of GSM MM are location registration with the PLMN and location updating to report the current location of a MS, as well as the identification and authentication of subscribers. These actions are closely interrelated. During registration into a GSM network, during the location updating procedure, and also during the setup of a connection, the identity of a mobile subscriber must be determined and verified (authentication).

The MM data are the foundation for creating the functions needed for routing and switching of user connections and for the associated services. For example, they are requested for routing an incoming call to the current MSC or for localizing a MS before paging is started. In addition to mobility data management, information about the configuration of SS is requested or changed, e.g. the currently valid target number for unconditional call forwarding in the HLR or VLR registers.

GSM – Architecture, Protocols and Services Third Edition   J. Eberspächer, H.-J. Vögel, C. Bettstetter and C. Hartmann
© 2009 John Wiley & Sons, Ltd

6.2 Location registration and location update

Before a MS can be called or gain access to services, the subscriber has to register with the mobile network (PLMN). This is usually the home network where the subscriber has a service contract. However, the subscriber can equally register with a foreign network provider in whose service area they are currently visiting, provided that there is a roaming agreement between the two network operators. Registration is only required if there is a change of networks, and therefore a VLR of the current network has not yet issued a TMSI to the subscriber. This means the subscriber has to report to the current network with their IMSI and receives a new TMSI by executing a location registration procedure. This TMSI is stored by the MS in its nonvolatile SIM storage, such that even after a powerdown and subsequent power-up only a normal location updating procedure is required.

The sequence of operations for registration is presented schematically in Figure 6.1. After a subscriber has requested registration at their current location by sending a LOCATION UPDATE REQUEST with their IMSI and the current location area (LAI), first the MSC instructs the VLR with a MAP message UPDATE LOCATION AREA to register the MS with its current LAI. In order for this registration to be valid, the identity of the subscriber has to be checked first, i.e. the authentication procedure is executed. For this purpose, the authentication parameters have to be requested from the AUC through the HLR. The precalculated sets of security parameters (Kc, RAND, SRES) are not usually transmitted individually to the respective VLR. In most cases, several complete sets are kept at hand for several authentications. Each set of parameters, however, can only be used once, i.e. the VLR must continually update its supply of security parameters (AUTHENTICATION PARAMETER REQUEST).

After successful authentication (section 5.6.2), the subscriber is assigned a new MSRN, which is stored with the LAI in the HLR, and a new TMSI is also reserved for this subscriber; this is TMSI reallocation (Figure 5.25). To encrypt the user data, the base station needs the ciphering key Kc, which it receives from the VLR by way of the MSC with the command START CIPHERING. After ciphering of the user data has begun, the TMSI is sent in encrypted form to the MS. Simultaneously with the TMSI assignment, the correct and successful registration into the PLMN is acknowledged (LOCAPDATE ACCEPT). Finally, the MS acknowledges the correct reception of the TMSI (TMSI REALLOCATION COMPLETE; Figure 5.26).

While the location information is being updated, the VLR is obtaining additional information about the subscriber, e.g. the MS category or configuration parameters for SS. For this purpose, the insert subscriber data procedure is defined (INSERT SUBSCRIBER DATA message in Figure 6.1). It is used for registration or location updating in the HLR to transmit the current data of the subscriber profile to the VLR. In general, this MAP procedure can always be used when the profile parameters are changed, e.g. if the subscriber reconfigures a SS such as unconditional forwarding. The changes are communicated immediately to the VLR with the insert subscriber data procedure.

The location update procedure is executed if the MS recognizes (by reading the LAI broadcast on the BCCH) that it is in a new location area, which leads to updating the location information in the HLR record. Alternatively, the location update can also occur periodically, independent of the current location. For this purpose, a time interval value is broadcast on the BCCH, which prescribes the time between location updates. The main objective of this

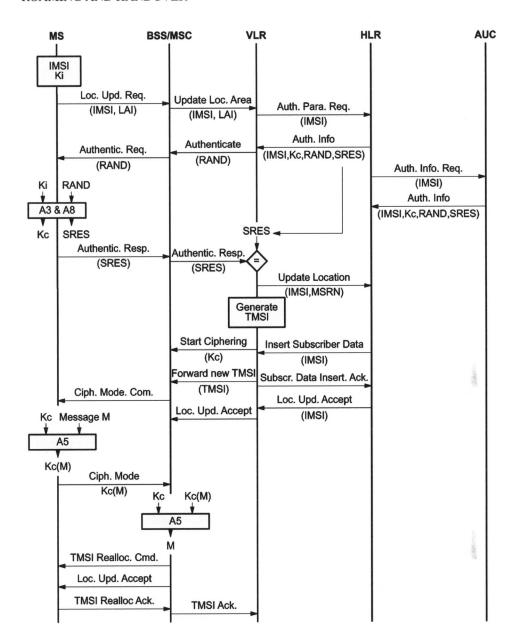

Figure 6.1 Overview of the location registration procedure.

location update is to know the current location for incoming calls or short messages, so that the call or message can be directed to the current location of the MS. The difference between the location update procedure and the location registration procedure is that in the first case

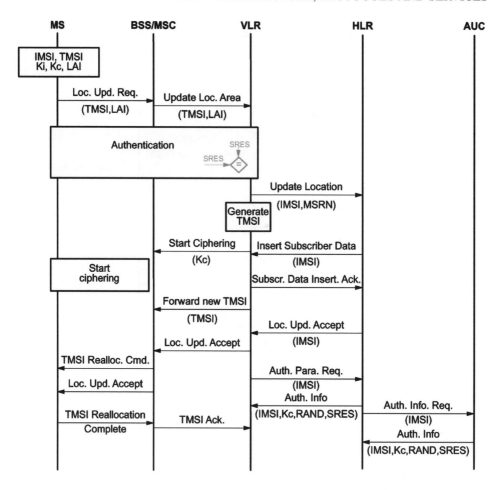

Figure 6.2 Overview of the location updating procedure.

the MS has already been assigned a TMSI. The TMSI is unique only in connection with an LAI, and both are kept together in the nonvolatile storage of the SIM card. With a valid TMSI, the MS also keeps a current ciphering key Kc for encryption of user data (Figure 6.2), although this key is renewed during the location update procedure. This key is recalculated by the MS based on the random number RAND used for authentication, whereas on the network side it is calculated in the AUC and made available in the VLR.

Corresponding to the location update procedure, there is a MM procedure at the air interface of the MM category 'specific'. In addition to the location updating proper, it contains three blocks which are realized at the air interface by three procedures of the category 'common' (Figure 5.26): the identification of the subscriber, the authentication and the start of ciphering on the radio channel. In the course of location updating, the MS also receives a new TMSI, and the current location is updated in the HLR. Figure 6.2 illustrates the standard case of a location update. The MS has entered a new LA, or the timer for periodic

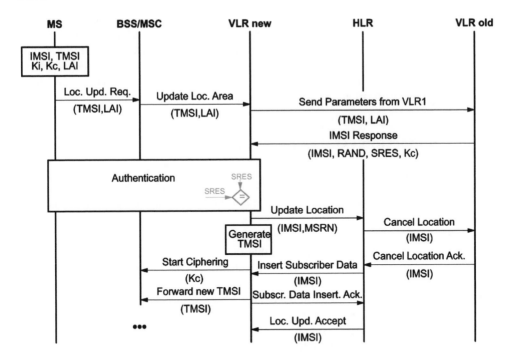

Figure 6.3 Location update after changing the VLR area.

location updating has expired, and the MS requests to update its location information. It is assumed that the new LA still belongs to the same VLR as the previous LA, so only a new TMSI needs to be assigned. This is the most frequent case. However, if it is not quite so crucial to keep the subscriber identity confidential, it is possible to avoid assigning a new TMSI. In this case, only the location information is updated in the HLR/VLR.

The new TMSI is transmitted to the MS in enciphered form together with the acknowledgement of the successful location update. The location update is complete after acknowledgement by the MS. After execution of the authentication, the VLR can complete its database and replace the 'consumed' 3-tuple (RAND, SRES, Kc) by another one requested from the HLR/AUC.

If location change involves both LA and VLR, the location update procedure is somewhat more complicated (Figure 6.3). In this case, the new VLR has to request the identification and security data for the MS from the old VLR and store them locally. Only in emergency cases, if the old VLR cannot be determined from the old LAI or if the old TMSI is not known in the VLR, the new VLR may request the IMSI directly from the MS (identification procedure). Only after a MS has been identified through the IMSI from the old VLR and after the security parameters are available in the new VLR, is it possible for the MS to be authenticated and registered in the new VLR, for a new TMSI to be assigned, and for the location information in the HLR to be actualized. After successful registration in the new VLR (LOCATION UPDATE ACCEPT) the HLR instructs the old VLR to cancel the invalid location data in the old VLR (CANCEL LOCATION).

In the examples shown (Figures 6.1–6.3), the location information is stored as a MSRN in the HLR. The MSRN contains the routing information for incoming calls and this information is used to route incoming calls to the current MSC. In this case, the HLR receives the routing information already at the time of the location update. Alternatively, at location update time, the HLR may just store the current MSC and/or VLR number in connection with an LMSI, such that routing information is only determined at the time of an incoming call.

6.3   Connection establishment and termination

6.3.1   Routing calls to MSs

The number dialed to reach a mobile subscriber (MSISDN) contains no information at all about the current location of the subscriber. In order to establish a complete connection to a mobile subscriber, however, one must determine the current location and the locally responsible switch (MSC). In order to be able to route the call to this switch, the routing address to this subscriber (MSRN) has to be obtained. This routing address is assigned temporarily to a subscriber by its currently associated VLR. At the arrival of a call at the GMSC, the HLR is the only entity in the GSM network which can supply this information, and therefore it must be interrogated for each connection setup to a mobile subscriber. The principal sequence of operations for routing to a mobile subscriber is shown in Figure 6.4. An ISDN switch recognizes from the MSISDN that the called subscriber is a mobile subscriber, and therefore can forward the call to the GMSC of the subscriber's home PLMN based on the CC and NDC in the MSISDN (1). This GMSC can now request the current routing address (MSRN) for the mobile subscriber from the HLR using the MAP (2,3). By way of the MSRN the call is forwarded to the local MSC (4), which determines the TMSI of the subscriber (5,6) and initiates the paging procedure in the relevant location area (7). After the MS has responded to the paging call (8), the connection can be switched through.

Several variants for determining the route and interrogating the HLR exist, depending on how the MSRN was assigned and stored, whether the call is national or international and depending on the capabilities of the associated switching centers.

Effect of the MSRN assignment on routing

There are two ways to obtain the MSRN:

- obtaining the MSRN at location update;

- obtaining the MSRN on a per call basis.

For the first variant, an MSRN for the MS is assigned at the time of each location update which is stored in the HLR. In this way the HLR is in a position to immediately supply the routing information needed to switch a call through to the local MSC.

The second variant requires that the HLR has at least an identification for the currently responsible VLR. In this case, when routing information is requested from the HLR, the HLR first has to obtain the MSRN from the VLR. This MSRN is assigned on a per call basis, i.e. each call involves a new MSRN assignment.

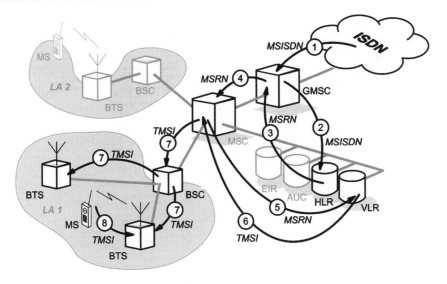

Figure 6.4 Principle of routing calls to mobile subscribers.

Placement of the protocol entities for HLR interrogation

Depending on the capabilities of the associated switches and the called target (national or international MSISDN), there are different routing procedures. In general, the local switching center analyzes the MSISDN. Owing to the NDC, this analysis of the MSISDN allows the separation of the mobile traffic from other traffic. The case that mobile call numbers are integrated into the numbering plan of the fixed network is currently not provided.

In the case of a national number, the local exchange recognizes from the NDC that the number is a mobile ISDN number. The fixed network and home PLMN of the called subscriber reside in the same country. In the ideal case, the local switch can interrogate the HLR responsible for this MSISDN (HLR in the home PLMN of the subscriber) and obtain the routing information (Figure 6.5(a)). The connection can then be switched through via fixed connections of the ISDN directly to the MSC.

If the local exchange does not have the required protocol intelligence for the interrogation of the HLR, the connection can be passed on preliminarily to a transit exchange, which then assumes the HLR interrogation and routing determination to the current MSC (Figure 6.5(b)). If the fixed network is not at all capable of performing a HLR interrogation, the connection has to be directed through a GMSC. This GMSC connects through to the current MSC (Figure 6.5(c)). For all three cases, the MS could also reside in a foreign PLMN (roaming); the connection is then made through international lines to the current MSC after interrogating the HLR of the home PLMN.

In the case of an international call number, the local exchange recognizes only the international CC and directs the call to an ISC. Then the ISC can recognize the NDC of the mobile network and process the call accordingly. Figures 6.6 and 6.7 show examples for the processing of routing information. An international call to a mobile subscriber involves at least three networks: the country from which the call originates; the country with the home

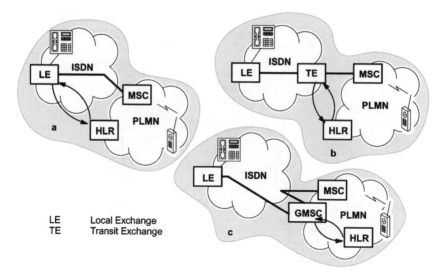

Figure 6.5 Routing variants for national MSISDN.

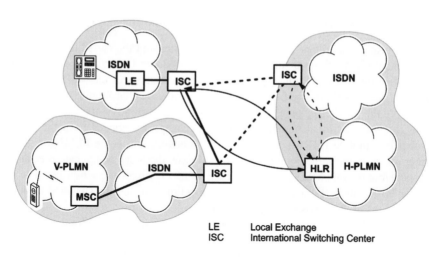

Figure 6.6 Routing for international MSISDN (HLR interrogation from ISC).

PLMN of the subscriber, Home PLMN (H-PLMN); and the country in which the mobile subscriber is currently roaming, Visited PLMN (V-PLMN). The traffic between countries is routed through ISCs. Depending on the capabilities of the ISC, there are several routing variants for international calls to mobile subscribers. The difference is determined by the entity that performs the HLR interrogation, resulting in differently occupied line capacities.

If the ISC performs the HLR interrogation, the routing to the current MSC is performed either by the ISC of the originating call or by the ISC of the mobile subscriber's H-PLMN

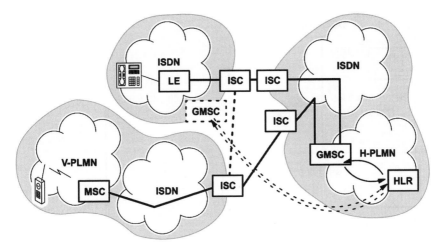

Figure 6.7 Routing through GMSC for international MSISDN.

(Figure 6.6). If no ISC can process the routing, again a GMSC has to become involved, either a GMSC in the country where the call originates or the GMSC of the H-PLMN (Figure 6.7).

For the routing procedures explained here, it does not matter which kind of subscriber is calling, i.e. the subscriber may be in the fixed network or in the mobile network. However, for calls from mobile subscribers, the HLR interrogation is usually performed at the local exchange (MSC).

6.3.2 Call establishment and corresponding MAP procedures

Call establishment in GSM at the air interface is similar to ISDN call establishment at the user network interface (Q.931) (Bocker, 1990). The procedure is supplemented by several functions: random access to establish a signaling channel (SDCCH) for call setup signaling, the authentication part, the start of ciphering and the assignment of a radio channel.

The establishment of a connection always contains a verification of user identity (authentication) independent of whether it is a mobile-originated call setup or a mobile-terminated call setup. The authentication is performed in the same way as for location updating. The VLR supplements its database entry for this MS with a set of security data, which replaces the 'consumed' 3-tuple (RAND, SRES, Kc). After successful authentication, the ciphering process for the encryption of user data is started.

Outgoing connection setup

For outgoing connection setup (Figure 6.8), first the MS announces its connection request to the MSC with a SETUP INDICATION message, which is a pseudo-message. It is generated between the MM entity of the MSC and the MAP entity, when the MSC receives the message CM-SERVICE REQUEST from the MS, which indicates in this way the request for a MM connection (Figure 5.27). Then the MSC signals to the VLR that the MS identified by the temporary TMSI in the location area LAI has requested service access (PROCESS ACCESS

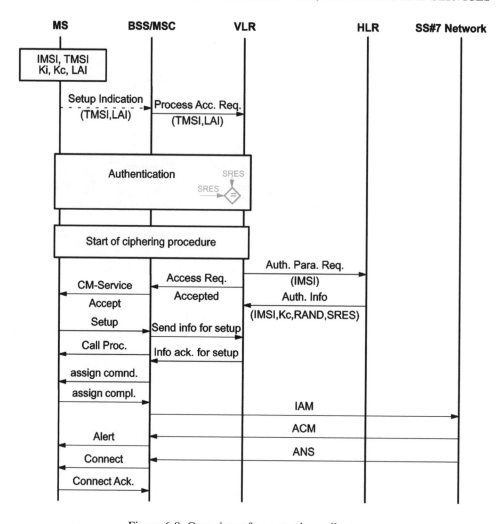

Figure 6.8 Overview of an outgoing call setup.

REQUEST) which is an implicit request for a random number RAND from the VLR, to be able to start the authentication of the MS. This random number is transmitted to the MS, its response with authentication result SRES is returned to the VLR, which now examines the authenticity of the MS's identity (compare authentication at registration; Figure 6.1).

After successful authentication, the ciphering process is started on the air interface, and this way the MM connection between MS and MSC has been completely established (CM-SERVICE ACCEPT). Subsequently, all signaling messages can be sent in encrypted form. Only now the MS reports the desired calling target. While the MS is informed with a CALL PROCEEDING message that processing of its connection request has been started, the MSC reserves a channel for the conversation and assigns it to the MS (ASSIGN). The connection request is signaled to the remote network exchange through the signaling system

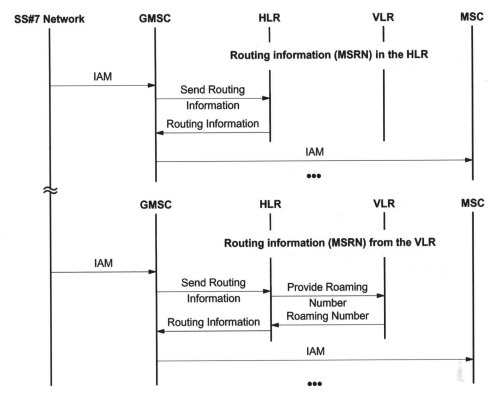

Figure 6.9 Interrogation of routing information for an incoming call.

SS#7 with the ISUP message IAM (Bocker, 1990). When the remote exchange answers (ACM), the delivery of the call can be indicated to the MS (ALERT). Finally, when the called partner goes off-hook, the connection can be switched through (CONNECT, ANS, CONNECT ACKNOWLEDGE).

Incoming connection setup

For incoming connection setup, it is necessary to determine the exact location of a MS in order to route the call to the currently responsible MSC. A call to a MS is therefore always routed to an entity which is able to interrogate the HLR for temporary routing information and to use it to forward the call. Usually, this entity is a GMSC of the home PLMN of the MS (section 6.3.1). Through this HLR interrogation, the GMSC obtains the current MSRN of the MS and forwards it to the current MSC (Figure 6.9).

Depending on whether the MSRN is stored in the HLR or first has to be obtained from the VLR, two variants of the HLR interrogation exist. In the first case, the interrogated HLR can supply the MSRN immediately (ROUTING INFORMATION). In the second case, the HLR has only received and stored the current VLR address during location update. Therefore, the HLR first has to request the current routing information from the VLR before the call can be switched through to the local MSC.

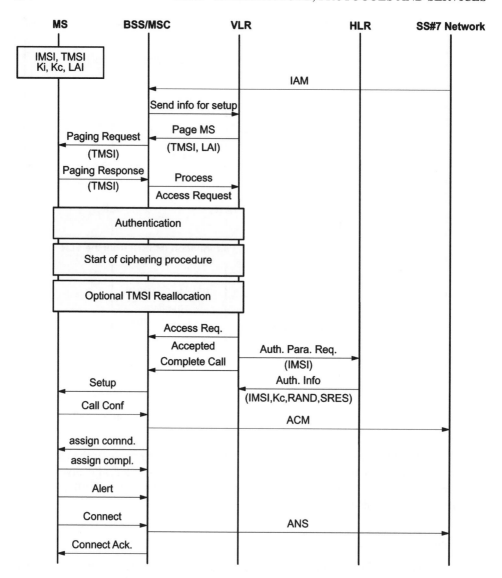

Figure 6.10 Overview of an incoming call setup.

Call processing is interrupted again in the local MSC in order to determine the exact location of the MS within the MSC area (SEND INFO FOR SETUP, Figure 6.10). The current LAI is stored in the location registers, but an LA can comprise several cells. Therefore, a broadcast (paging call) in all cells of the LA is used to determine the exact location, i.e. cell, of the MS. Paging is initiated from the VLR using the MAP (PAGE MS) and transformed by the MSC into the paging procedure at the air interface. When a MS receives a paging call, it responds directly and thus allows the current cell to be determined.

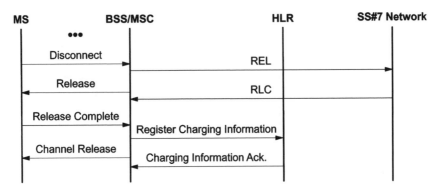

Figure 6.11 Mobile-initiated call termination and storing of charging information.

Thereafter, the VLR instructs the MSC to authenticate the MS and to start ciphering on the signaling channel. Optionally, the VLR can execute a reallocation of the TMSI (TMSI reallocation procedure) during call setup. Only at this point, after the network internal connection has been established (section 5.4.4), can the connection setup proper be processed (command COMPLETE CALL from VLR to MSC). The MS is told about the connection request with a SETUP message, and after answering CALL COMPLETE it receives a channel. After ringing (ALERT) and going off-hook, the connection is switched through (CONNECT, CONNECT, ACKNOWLEDGE), and this fact is also signaled to the remote exchange (ACM, ANS).

6.3.3 Call termination

At the air interface, a given call can be terminated either by the mobile equipment or by the network. The taking down of the connection is initiated at the Um interface by means of the CC messages DISCONNECT, RELEASE and RELEASE COMPLETE. This is followed by an explicit release of occupied radio resources (CHANNEL RELEASE). On the network side, the connection between the involved switching centers (MSC, etc.) is terminated using the ISUP messages REL and RLC in the SS#7 network (Figure 6.11).

After the connection has been taken down, information about charges (CHARGING INFORMATION) is stored in the VLR or HLR using the MAP. This charging data can also be required for an incoming call, e.g. if roaming charges are due because the called subscriber is not in their home PLMN.

6.3.4 MAP procedures and routing for short messages

A connectionless relay protocol has been defined for the transport of short messages (section 5.4.8) at the air interface, which has a counterpart in the network in a store-and-forward operation for short messages. This forwarding of transport PDUs of the SMS uses MAP procedures. For an incoming short message which arrives from the SMS-SC at a SMS-GMSC, the exact location of the MS is the first item that needs to be determined just as for an incoming call. The current MSC of the MS is first obtained with a HLR

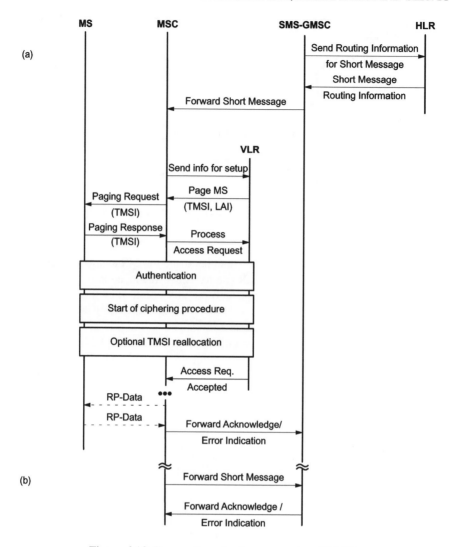

Figure 6.12 Forwarding short messages in a PLMN.

interrogation (SHORT MESSAGE ROUTING INFORMATION; Figure 6.12(a)). The short message is then passed to this MSC (FORWARD SHORT MESSAGE) and is locally delivered after paging and SMS connection setup. Success or failure are reported to the SMS-GMSC in another MAP message (FORWARD ACKNOWLEDGEMENT/ERROR INDICATION) which then informs the service center.

In the reverse case, for an outgoing short message, no routing interrogation is needed, since the SMS-GMSC is known to all MSCs, so the message can be passed immediately to the SMS-GMSC (Figure 6.12(b)).

6.4 Handover

6.4.1 Overview

Handover is the transfer of an existing voice connection to a new base station. There are different reasons for the handover to become necessary. In GSM, a handover decision is made by the network, not the MS, and it is based on BSS criteria (received signal level, channel quality, distance between MS and BTS) and on network operation criteria (e.g. current traffic load of the cell and ongoing maintenance work).

The functions for preparation of handover are part of the radio subsystem link control. Above all, this includes the measurement of the channel. Periodically, a MS checks the signal field strength of its current downlinks as well as those of the neighboring base stations, including their BSICs. The MS sends measurement reports to its current base station (quality monitoring); section 4.5.1. On the network side, the signal quality of the uplink is monitored, the measurement reports are evaluated, and handover decisions are made.

As a matter of principle, handovers are only performed between base stations of the same PLMN. Handovers between BSSs in different networks are not allowed. Two kinds of handover are distinguished (Figure 6.13).

- *Intracell handover*: for administrative reasons or because of channel quality (channel-selective interferences), the MS is assigned a new channel within the same cell. This decision is made locally by the RR of the BSS and is also executed within the BSS.

- *Intercell handover*: the connection to a MS is transferred over the cell boundary to a new BTS. The decision about the time of handover is made by the RR protocol module of the network based on measurement data from MSs and BSSs. The MSC, however, can participate in the selection of the new cell or BTS. The intercell handover occurs most often when it is recognized from weak signal field strength and bad channel quality (high bit error ratio) that a MS is moving near the cell boundary. However, an intercell handover can also occur due to administrative reasons, say for traffic load balancing. The decision about such a network-directed handover is made by the MSC, which instructs the BSS to select candidates for such a handover.

Two cases need to be distinguished between with regard to participation of network components in the handover, depending on whether the signaling sequences of a handover execution also involve a MSC. Since the RR module of the network resides in the BSC (see Figure 5.11), the BSS can perform the handover without participation of the MSC. Such handovers occur between cells which are controlled by the same BSC and are called internal handover. They can be performed independently by the BSS; the MSC is only informed about the successful execution of internal handovers. All other handovers require participation of at least one MSC, or their BSSMAP and MAP parts, respectively. These handovers are known as external handovers.

Participating MSCs can act in the role of MSC-A or MSC-B. MSC-A is the MSC which performed the initial connection setup, and it keeps the MSC-A role and complete control (anchor MSC) for the entire life of the connection. A handover is therefore, in general, the extension of the connection from the anchor MSC-A to another MSC (MSC-B). In this case, the mobile connection is passed from MSC-A to MSC-B with MSC-A keeping the ultimate control over the connection. An example is presented in Figure 6.14. A MS occupies an

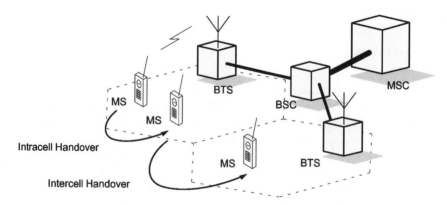

Figure 6.13 Intracell and intercell handover.

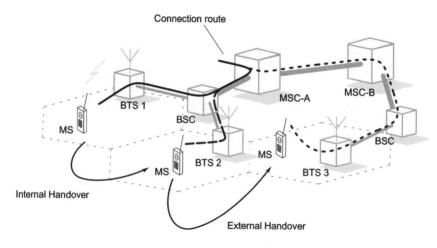

Figure 6.14 Internal and external handover.

active connection via BTS1 and moves into the next cell. This cell of BTS2 is controlled by the same BSC so that an internal handover is indicated. The connection is now carried from MSC-A over the BSC and the BTS2 to the MS; the connections of BTS1 (radio channel and ISDN channel between BTS and BSC) were taken down. As the MS moves on to the cell handled by BTS3, it enters a new BSS which requires an external handover. In addition, this BSS belongs to another MSC, which now has to play the role of MSC-B. Logically, the connection is extended from MSC-A to MSC-B and carried over the BSS to the MS. At the next change of the MSC, the connection element between MSC-A and MSC-B is taken down, and a connection to the new MSC from MSC-A is set up. Then the new MSC takes over the role of MSC-B.

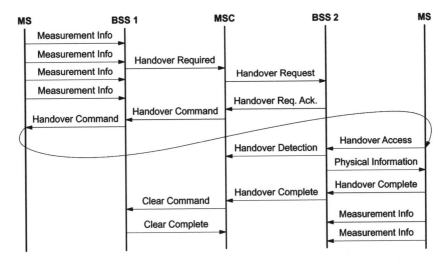

Figure 6.15 Principal signaling sequence for an intra-MSC handover.

6.4.2 Intra-MSC handover

The basic structure for an external handover is the handover between two cells of the same MSC (Figure 6.15). The MS continually transmits measurement reports with channel monitoring data on its SACCH to the current base station (BSS 1). Based on these measurement results, the BSS decides when to perform a handover and requests this handover from the MSC (message HANDOVER REQUIRED). The respective measurement results can be transmitted in this message to the MSC, to enable its participation in the handover decision. The MSC causes the new BSS to prepare a channel for the handover, and frees the handover to the MS (HANDOVER COMMAND), as soon as the reservation is acknowledged by the new BSS. The MS now reports to the new BSS (HANDOVER ACCESS) and receives information about the properties of the physical channel. This includes synchronization data such as the new timing advance value and also the new transmitter power level. Once the MS is able to occupy the channel successfully, it acknowledges this fact with a message HANDOVER COMPLETE. The resources of the old BSS can then be released.

6.4.3 Decision algorithm for handover timing

The basis for processing a successful handover is a decision algorithm which uses measurement results from the MS and base station to identify possible other base stations as targets for handovers and which determines the optimal moment to execute the handover. The objective is to keep the number of handovers per cell change as low as possible. Ideally, there should not be more than one handover per cell change. In reality, this is often not achievable. When a MS leaves the radio range of a base station and enters one of a neighboring station, the radio conditions are often not very stable, so that several handovers must be executed before a stable state is reached. Simulation results of Mende (1991) and Junius and Marger (1992) give a mean value of about 1.5–5 handovers per cell change.

Since every handover incurs not only increased traffic load for the signaling and transport system but also reductions in speech quality, the importance of a well-dimensioned handover decision algorithm is obvious, an algorithm which also takes into account the momentary local conditions. This is also a reason for the lack of a standardized, uniform algorithm for the determination of the moment of the handover in GSM. For this decision about when to perform a handover, network operators can develop and deploy their own algorithms which are optimally tuned for their networks. This is made possible through standardizing only the signaling interface that defines the processing of the handover and through transferring the handover decision to the BSS. The GSM handover is thus a network-originated handover as opposed to a mobile-originated handover, where the handover decision is made by the MS. An advantage of this handover approach is that the software of the MS need not be changed when the handover strategy or the handover decision algorithm is changed in all or part of the network. Even though the GSM standard does not prescribe a mandatory handover decision algorithm, a simple algorithm is proposed, which can be selected by the network operator or replaced by a more complex algorithm.

In principle, a GSM handover always proceeds in three steps (Figure 6.16), which are based on the measurement data provided by the MS over the SACCH, and on the measurements performed by the BSS itself. Foremost among these data items are the current channel's received signal level (RXLEV) and the signal quality (RXQUAL), both on the uplink (measured by the BSS) and on the downlink (measured by the MS). In order to be able to identify neighboring cells as potential targets for a handover, the MS also measures the received signal level RXLEV_CELL(n) of up to 16 neighboring base stations. The RXLEV values of the six base stations which can be received best are reported every 480 ms to the BSS. Further criteria for the handover decision algorithm are the distance between MS and BTS measured via the TA of the adaptive frame alignment (section 4.3.2) and measurements of the interference in unused time slots. A new value of each of these measurements is available every 480 ms.

Measurement preprocessing calculates average values from these measurements, whereby at least the last 32 values of RXLEV and RXQUAL must be averaged. The resulting mean values are continuously compared with thresholds (see Table 6.1) after every SACCH interval.

These threshold values can be configured individually for each BSS through management interfaces of the OMSS (see Chapter 3). The principle used for the comparison of measurements with the threshold is to conduct a so-called Bernoulli experiment: if out of the last N_i mean values of a criterion i more than P_i go under (RXLEV) or over (RXQUAL, MS_RANGE) the threshold, then a handover may be a necessary. The values of N_i and P_i can also be configured through network management. Their allowed range is defined as the interval [0; 31].

In addition to these mean values, a BSS can calculate the current power budget (PBGT(n)), which represents a measure for the respective path loss between MS and current base station or a neighboring base station n. Using this criterion, a handover can always be caused to occur to the base station with the least path loss for the signals from or to the MS. The PBGT takes into consideration not only the RXLEV_DL of the current downlink and the RXLEV_NCELL(n) of the neighboring BCCH but also the maximal transmitter power P (see Table 4.8) of a MS, the maximal power MS_TXPWR_MAX allowed to a MS in the current cell, and the maximal power MS_TXPWR_MAX(n) allowed to mobiles in

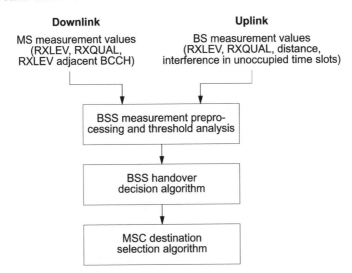

Figure 6.16 Decision steps in a GSM handover.

the neighboring cells. In addition, the calculation uses the value PWR_C_D, which is the difference between maximal transmitter power on the downlink and current transmitter power of the BTS in the downlink, a measure for the available power control reserve.

Thus, the power budget for a neighboring base station n is calculated as follows:

$$PBGT(n) = (Minimum(MS\_TXPWR\_MAX, P) - RXLEV\_DL - PWR\_C\_D)$$
$$- (Minimum(MS\_TXPWR\_MAX(n), P) - RXLEV\_NCELL(n)).$$

A handover to a neighboring base station can be requested, if the power budget is $PBGT(n) > 0$ and greater than the threshold HO_MARGIN(n). The causes for handover which are possible using these criteria are summarized in Table 6.2. As can be seen, the signal criteria of the uplink and downlink as well as the distance from the base station and PBGT can lead to a handover.

The BSS makes a handover decision by first determining the necessity of a handover using the threshold values of Table 6.1. In principle, one can distinguish three categories:

- handover because of more favorable path loss conditions;

- mandatory intercell handover;

- mandatory intracell handover.

Situations where a neighboring base station shows more favorable propagation conditions and therefore lower path loss, do not necessarily force a handover. Such potential handover situations to a neighboring cell are discovered through the PBGT(n) calculations. To make a handover necessary, the power budget of the neighboring cell must be greater than the threshold HO_MARGIN(n).

Table 6.1 Threshold values for the GSM handover.

Threshold value	Typical value	Meaning
L_RXLEV_UL_H	103 to 73 dBm	Upper handover threshold of the received signal level in the uplink
L_RXLEV_DL_H	103 to 73 dBm	Upper handover threshold of the received signal level in the downlink
L_RXLEV_UL_IH	85 to 40 dBm	Lower(!) received signal level threshold in the uplink for internal handover
L_RXLEV_DL_IH	85 to 40 dBm	Lower(!) received signal level threshold in the downlink for internal handover
RXLEV_MIN(n)	Approximately 85 dBm	Minimum required RXLEV of BCCH of cell n to perform a handover to this cell
L_RXQUAL_UL_H	–	Lower handover threshold of the bit error ratio in the uplink
L_RXQUAL_DL_H	–	Lower handover threshold of the bit error ratio in the downlink
MS_RANGE_MAX	2 to 35 km	Maximum distance between the MS and base station
HO_MARGIN(n)	0 to 24 dB	Hysteresis to avoid multiple handovers between two cells

Table 6.2 Handover causes.

Handover cause	Meaning
UL_RXLEV	Uplink received signal level too low
DL_RXLEV	Downlink received signal level too low
UL_RXQUAL	Uplink bit error ratio too high
DL_RXQUAL	Downlink bit error ratio too high
PWR_CTRL_FAIL	Power control range exceeded

The recognition of a mandatory handover situation (Figure 6.17) within the framework of the radio subsystem link control (section 4.5 and Figure 4.19) is based on the received signal level and signal quality in uplink and downlink as well as on the distance between the MS and BTS. Going over or under the respective thresholds always necessitates a handover. Here are the typical situations for a mandatory handover:

- The received signal level in the uplink or downlink (RXLEV_UL/RXLEV_DL) drops below the respective handover threshold value (L_RXLEV_UL_H/L_RXLEV _DL_H) and the power control range has been exhausted, i.e. the MS and/or the BSS have reached their maximal transmitter power (section 4.5.2).

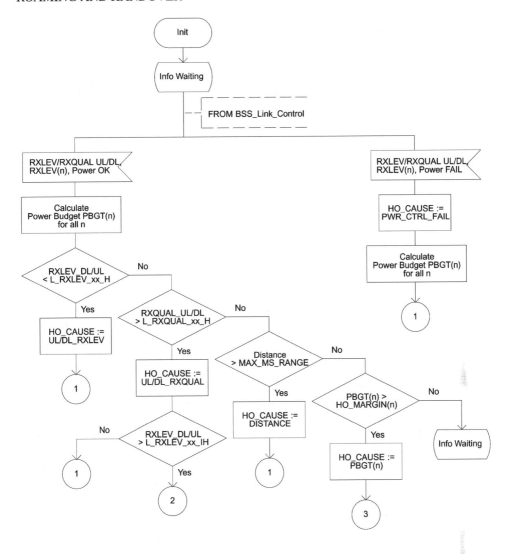

Figure 6.17 Detection of mandatory handover.

- The bit error ratio as a measure of signal quality in uplink and/or downlink
  (RXQUAL_UL/RXQUAL_DL)
  exceeds the respective handover threshold value
  (L_RXQUAL_UL_H/L_RXQUAL_DL_H),
  while at the same time the received signal level drops into the neighborhood of the
  threshold value.

- The maximum distance to the base station (MAX_MS_RANGE) has been reached.

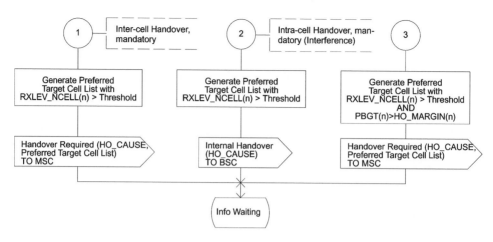

Figure 6.18 Completion of the handover decision in the BSS.

A handover can also become mandatory when the handover thresholds are not exceeded if the lower thresholds of the transmitter power control are exceeded (L_RXLEV_ xx_P/L_RXQUAL_xx_P; Table 4.9), even though the maximum transmitter power has already been reached. The cause of handover indicated is the failure of the transmitter power control (PWR_CTR_FAIL; Table 6.2).

A special handover situation exists if the bit error ratio RXQUAL as a measurement for signal quality in the uplink and/or downlink exceeds its threshold and at the same time the received signal level is greater than the thresholds L_RXLEC_UL_IH/L_ RXLEC_DL_IH. This strongly hints at an existing severe cochannel interference. This problem can be solved with an (internal) intracell handover, which the BSS can perform on its own without support from the MSC. It is also considered as a mandatory handover.

If the BSS has detected a handover situation, a list of candidates as possible handover targets is assembled using the BSS decision algorithm. For this purpose, one first determines which BCCH of the neighboring cell n is received with sufficient signal level:

$$\text{RXLEV\_NCELL}(n) > \text{RXLEV\_MIN}(n) + \text{Maximum}(0, (\text{MS\_TXPWR\_MAX}(n) - P))).$$

The potential handover targets are then assembled in an ordered list of preferred cells according to their path loss compared with the current cell (Figure 6.18). For this purpose, the power budget of the neighboring cells in question is again evaluated:

$$\text{PBGT}(n) - \text{HO\_MARGIN}(n) > 0.$$

All cells n which are potential targets for a handover due to RXLEV_NCELL(n) and lower path loss than the current channel are then reported to the MSC with the message HANDOVER REQUIRED (Figure 6.18) as possible handover targets. This list is sorted by priority according to the difference (PBGT(n) − HO_MARGIN(n)). The same message HANDOVER REQUIRED is also generated if the MSC has sent a message HANDOVER CANDIDATE ENQUIRY to the BSS.

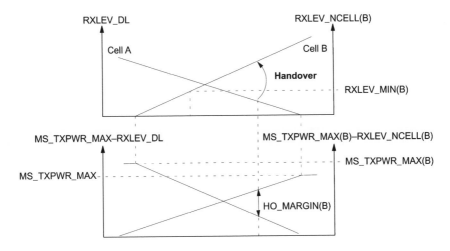

Figure 6.19 Handover criteria for exhausted transmitter power control.

The conditions at a cell boundary in the case of exhausted transmitter power control (PWR_C_D = 0) are shown in Figure 6.19 with a MS moving from the current cell to a cell B. The threshold RXLEV_MIN(B) is reached very early; however, the handover is somewhat moved in the direction of cell B because of the positive HO_MARGIN(B) for the power budget. When moving in the opposite direction, the handover would be delayed in the other direction due to HO_MARGIN(A) of cell A. This has the effect of a hysteresis which reduces repeated handovers between both cells due to fading (ping-pong handover). In addition to varying radio conditions (fading due to multipath propagation, shadowing, etc.) there are many other sources of error with this kind of handover. Note that, on the one hand, that there are substantial delays between measurement and reaction due to the averaging process. This leads to executing the handover too late on a few occasions. It is more important, however, that the current channel is compared with the BCCH of the neighboring cells rather than the traffic channel to be used after the handover decision, which could suffer from different propagation conditions (frequency-selective fading, etc.). Finally, the MSC decides about the target cell of the handover. This decision takes into consideration the following criteria in decreasing order of priority: handover due to signal quality (RXQUAL), received signal level (RXLEV), distance and path loss (PBGT). This prioritization is especially effective when there are not enough traffic channels available and handover requests are competing for the available channels.

The standard explicitly points out that all measurement results must be sent with the message HANDOVER REQUIRED to the MSC, so that in the end the option remains open to implement the complete handover decision algorithm in the MSC.

6.4.4 MAP and inter-MSC handover

The most general form of handover is the inter-MSC handover. The MS moves over a cell boundary and enters the area of responsibility of a new MSC. The handover caused by this

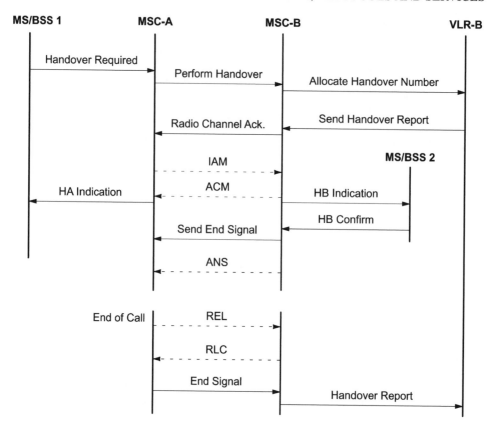

Figure 6.20 Principal operation of a basic handover.

move requires communication between the involved MSCs. This occurs through the SS#7 using transactions of the MAP.

Basic handover between two MSCs

The principal sequence of operations for a basic handover between two MSCs is shown in Figure 6.20. The MS has indicated the conditions for the handover, and the BSS requests the handover from MSC-A (HANDOVER REQUIRED). MSC-A decides positively for a handover and sends a message PERFORM HANDOVER to MSC-B. This message contains the necessary data to enable MSC-B to reserve a radio channel for the MS. Above all, it identifies the BSS which is to receive the connection. MSC-B assigns a handover number and tries to allocate a channel for the MS. If a channel is available, the response RADIO CHANNEL ACKNOWLEDGE contains the new MSRN to the MS and the designation of the new channel. If no channel is available, this is also reported to MSC-A which then terminates the handover procedure.

When a RADIO CHANNEL ACKNOWLEDGE is successful, an ISDN channel is switched through between the two MSCs (ISUP messages IAM and ACM), and both MSCs send

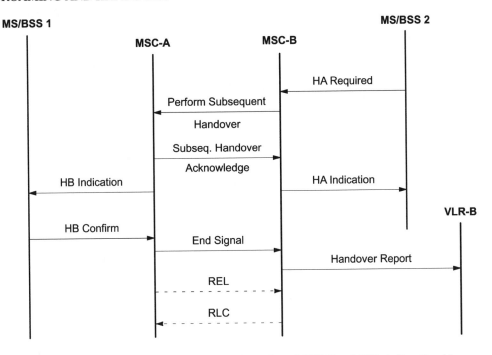

Figure 6.21 Principle of subsequent handover from MSC-B to MSC-A (handback).

an acknowledgement to the MS (HA INDICATION, HB INDICATION). The MS then resumes the connection on the new channel after a short interruption (HB CONFIRM). MSC-B then sends a message SEND END SIGNAL to MSC-A and thus causes the release of the old radio connection. After the end of the connection (ISUP messages REL, RLC), MSC-A generates a message END SIGNAL for MSC-B which then sends a HANDOVER REPORT to its VLR.

Subsequent handover

After a first basic handover of a connection from MSC-A to MSC-B, a MS can move on freely. Further intra-MSC handovers can occur (Figure 6.15), which are processed by MSC-B.

If, however, the MS leaves the area of MSC-B during this connection, a subsequent handover becomes necessary. Two cases are distinguished between: in the first case, the MS returns to the area of MSC-A, whereas in the second case it enters the area of a new MSC, now called MSC-B'. In both cases, the connection is newly routed from MSC-A. The connection between MSC-A and MSC-B is taken down after a successful subsequent handover.

A subsequent handover from MSC-B back to MSC-A is also called handback (Figure 6.21). In this case, MSC-A, as the controlling entity, does not need to assign a handover number and can search directly for a new radio channel for the MS. If a radio channel can be allocated in time, both MSCs start their handover procedures at the air interface (HA/HB INDICATION) and complete the handover. After completion, MSC-A

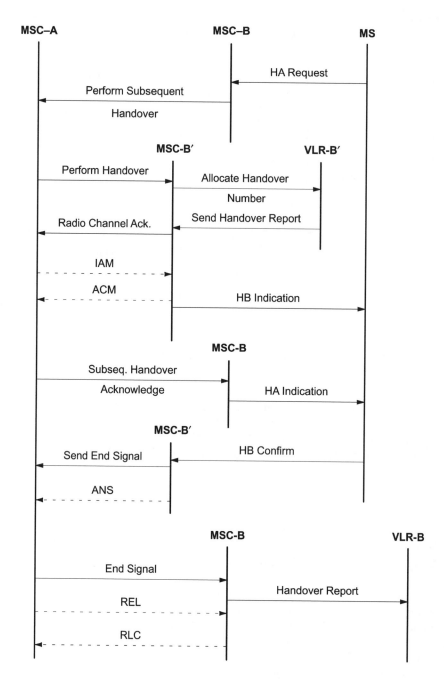

Figure 6.22 Principle of subsequent handover from MSC-B to MSC-B'.

terminates the connection to MSC-B. The message END SIGNAL terminates the MAP process in MSC-B and causes a HANDOVER REPORT to be sent to the VLR of MSC-B; the ISUP message RELEASE releases the ISDN connection.

The procedure for a subsequent handover from MSC-B to MSC-B is more complicated. It consists of two parts:

- a subsequent handover from MSC-B to MSC-A;

- a basic handover between MSC-A and MSC-B′.

The principal operation of this handover is illustrated in Figure 6.22.

In this case, MSC-A recognizes from the message PERFORM SUBSEQUENT HAN-DOVER, sent by MSC-B, that it is a case of handover to an MSC-B′, and it initiates a basic handover to MSC-B′. MSC-A informs MSC-B after receiving the ISUP message ACM from MSC-B about the start of handover at MSC-B′ and thereby frees the handover procedure at the radio interface from MSC-B. Once MSC-A receives the MAP message SEND END SIGNAL from MSC-B, it considers the handover as complete, sends the message END SIGNAL to MSC-B to terminate the MAP procedure and cancels the ISDN connection.

7

Services

In this section we summarize and describe the most important services provided in GSM systems.

7.1 Classical GSM services

7.1.1 Teleservices

Voice

Voice services had to be implemented by each operator in the start-up phase (E1) by 1991. In this category, two teleservices were distinguished: regular telephone service (TS11) and emergency service (TS12). For the transmission of the digitally coded speech signals, both services use a bidirectional, symmetric, full-duplex, point-to-point connection, which is set up on user demand. The sole difference between TS11 and TS12 teleservices is that regular service requires an international IWF, whereas the emergency service stays within the boundaries of a national network.

Fax transmission

As a teleservice for the second implementation phase (E2), the implementation of a transparent fax service (TS61) for Group 3 fax was planned. The fax service is called transparent because it uses a transparent bearer service for the transmission of fax data. The coding and transmission of the facsimile data uses the fax protocol according to the ITU-T recommendation T30. The network operator also has the option to implement TS61 on a nontransparent bearer service in order to improve the transmission quality. TS61 is transmitted over a traffic channel that is alternately used for voice or fax. Another optional alternative is designated as a fax transfer with automatic call acceptance (TS61). This service can be offered by a network operator when multinumbering is used as the interworking solution. In the case of multinumbering, a subscriber is assigned several MSISDN numbers, and a separate interworking profile is stored for each of them. In this way a specific teleservice

GSM – Architecture, Protocols and Services Third Edition   J. Eberspächer, H.-J. Vögel, C. Bettstetter and C. Hartmann
© 2009 John Wiley & Sons, Ltd

can be associated with each MSISDN, the fax service being one of them. If a mobile subscriber is called on their GSM-fax number, the required resources in the IWF of the MSC as well as in the MS can be activated; in the case of TS61, fax calls arrive with the same number as voice calls (no multinumbering) and have to be switched over to fax reception manually.

7.2 Popular GSM services: SMS and MMS

Next to voice services, which have been the main focus of GSM since the start, SMS has proven to be extremely popular and successful in GSM. To follow this success, similar but enhanced services have been standardized, such as the Enhanced Messaging Service (EMS) and the Multimedia Messaging Service (MMS). A brief description of these services follows in the next sections.

7.2.1 SMS

One of the most important services in GSM systems today, in terms of popularity at the user side as well as in terms of revenue generation at the provider side, is the capability to receive or send short messages at the MS: SMS, TS21 and TS22. This service was supposed to be offered in the third phase (E3) at the latest from 1996 on all GSM networks. TS21 is the point-to-point version of the SMS, which allows a single station to be sent a message of up to 160 characters. Conversely, TS22 has been defined as an optional implementation of the capability to send short messages from a MS. The combinations of SMS with other added-value services, e.g. mailbox systems with automatic notification of newly arrived messages or the transmission by short message of incurred charges, clearly show how the services offered by GSM networks go significantly beyond the services offered in fixed networks.

For SMS, the network operator has to establish a service center which accepts short messages from the fixed network and processes them in a store-and-forward mode. The interface has not been specified and can be by DTMF signaling, special order, email, fax, etc. The delivery can be time-shifted and is of course independent of the current location of the MS. Conversely, a service center can accept short messages from MSs which can also be forwarded to subscribers in the fixed network, for example by fax or email. The transmission of short messages uses a connectionless, protected, packet-switching protocol. The reception of a message must be acknowledged by the MS or the service center; in the case of failure, retransmission occurs. TS21 and TS22 are the only teleservices which can be used simultaneously with other services, i.e. short messages can also be received or transmitted during an ongoing call.

A further variation of the SMS is the cell broadcast service TS23, SMSCB. SMSCB messages are broadcast only in a limited region of the network. They can only be received by MSs in idle mode, and reception is not acknowledged. A MS itself cannot send SMSCB messages. With this service, messages contain a category designation, so that MSs can select categories of interest which they want to receive and store. The maximum length of SMSCB messages is 93 characters, but by using a special reassembly mechanism, the network can transmit longer messages of up to 15 subsequent SMSCB messages.

Table 7.1 SMS, EMS and MMS.

Service	Introduced	Payload size	Content
SMS	1995	160 byte	Text
EMS	2000	160 byte	Text, pictures, animations
MMS	2001	100 kbyte	Text, voice, pictures, photos, video

7.2.2 EMS

EMS is an extension of SMS. SMS was limited to text messages only. However, as ring tones and pictures gained a lot of popularity, EMS was introduced. EMS was developed by major GSM manufacturers as an open 3GPP standard. It allows unicolor pictures with 16×16 or 32×32 pixels to be sent and the pictures to be modified in the handset. Picture sequences can comprise six pictures. Fonts can be formatted in EMS and tones of three octaves can be included, from the pitch of C to the pitch of B++. The duration of tones can be 150, 225, 300 or 450 ms and up to 80 notes can be included in one EMS. However, before EMS had a chance to become popular, the MMS standard, which is much richer, took over.

7.2.3 MMS

MMS is similar to SMS or EMS, however has much higher capabilities in terms of size and flexibility. The MMS standard was developed by a consortium of industry partners and has become a 3GPP standard. In addition to pure text, MMS is capable of transmitting pictures, melodies and multimedia sequences of different kinds. MMS can transmit up to 100 kbyte of data and can handle AMR-coded speech, pictures (e.g. JPEG or GIF), music and even video. From the network point of view, a MMS-Center, called MMS-C, is required, which is responsible for storing, converting and forwarding MMS data. The MMS-C also stores information about the preferences of users as well as their terminal capabilities. Therefore, it is possible to avoid the transmission of MMS messages to a terminal which cannot deal with the specific format. Instead, the MMS message can possibly be transformed at the MMS-C into a format which the receiving terminal can handle. A comparison between SMS, EMS and MMS is provided in Table 7.1.

MMS architecture

The Multimedia Messaging Service Network Architecture (MMSNA) (Figure 7.1) comprises all required elements to provide a complete MMS to a user. This includes interworking between service providers. The Multimedia Messaging Service Environment (MMSE) is a collection of MMS-specific network elements which is controlled by a single administration. In the case of roaming, the visited network is part of the MMSE of the user in question, while subscribers of a different service provider are part of a different MMSE. An important role is taken on by the MMS relay/server, which is responsible for storing and handling both incoming and outgoing messages. It is also in charge of the transfer of messages between different messaging systems. The MMS relay/server can either be a single logical element

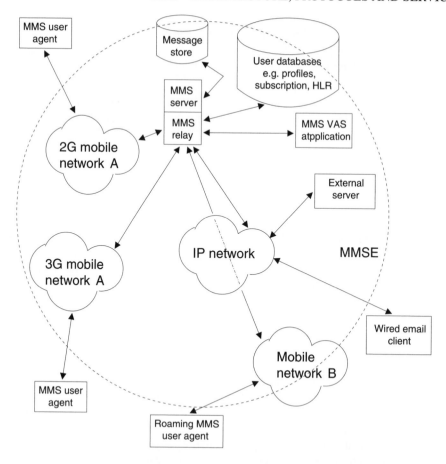

Figure 7.1 Multimedia Messaging Service Network Architecture (MMSNA).

or it may be separated into two single elements, a MMS relay and a MMS server. The MMS relay/server must also be able to generate charging data, when receiving Multimedia Messages (MMs) from or delivering MMs to other elements of the architecture. The MMS user database contains user-related information, such as subscription, configuration and capability data. The MMS user database can be either a single entity or it can be distributed. The MMS user agent is an application layer function, providing the users with the ability to view, compose and handle MMs. The MMS user agent resides on the mobile device. Finally, the MMS Value Added Services (VAS) applications offer value-added services to MMS users. Several MMS VAS applications may be connected to a MMSE.

7.3   Overview of GSM services in Phase 2+

The GSM standards have always been enhanced: Phase 1 of the GSM implementation contained basic teleservices – in the first place voice communication – and a few supplementary services, which had to be offered by all network operators in 1991 when GSM was introduced into the market. The standardization of Phase 2 was completed in 1995 with market introduction following in 1996. Essentially, ETSI added more of the supplementary services, which had been planned already when GSM was initially conceived and which were adopted from the fixed ISDN. These new services made it necessary to rework large parts of the GSM standards. For this reason, networks operating according to the revised standard are also called GSM Phase 2 (Mouly and Pautet, 1995). However, all networks and terminals of Phase 2 preserve the compatibility with the old terminals and network equipment of Phase 1, i.e. all new standard development had to be strictly backwards compatible.

The GSM Phase 2+ introduces many aspects ranging from radio transmission to communication and call processing. However, there is no complete revision of the GSM standard; rather single subject areas are treated as separate standardization units, with the intent of allowing them to be implemented and introduced independently of each other.

Thus, GSM systems can evolve gradually, and standardization can meet market needs in a flexible way. However, with this approach, a unique identification of a GSM standard version becomes impossible. The designation GSM Phase 2+ is supposed to indicate this openness (Mouly and Pautet, 1995), suggesting an evolutionary process with no endpoint in time or prescribed target dates for the introduction of new services. The GSM standards are now published in so-called releases (e.g. Release 97, 98, 99 and 2000).

A large menu of technical questions is being addressed, only a few of which are presented as examples in the following. Figure 7.2 illustrates the evolution of GSM, from the initial digital speech services towards the third generation of mobile communications (UMTS/IMT-2000). In particular, it shows the services of Phase 2+ that are covered in this book. Most of these services are already offered by GSM network providers today and can be used with enhanced mobile equipment.

7.4 Bearer and teleservices of GSM Phase 2+

Whereas GSM Phase 2 defined essentially a set of new supplementary services, Phase 2+ is also addressing new bearer and teleservices. In this section we give an overview of these new speech and data services. They significantly improve the GSM speech quality and make the utilization of available radio resources much more efficient. Furthermore, the new data services are an important step towards wireless Internet access via cellular networks.

7.4.1 Advanced speech call items

The GSM systems of Phase 2 offer inadequate features for group communications. For example, group call or 'push-to-talk' services with fast connection setup as known from private radio or digital trunked radio systems (e.g. TETRA) are not offered. However, such services are indispensable for most closed user groups (e.g. police, airport staff, railroad or taxi companies). In particular, railroad operators had a strong request for such features. In 1992, their international organization, the Union Internationale des Chemins de Fer (UIC), selected the GSM system as their standard (Mouly and Pautet, 1995). This GSM-based uniform international railway communication system should replace a multitude of incompatible radio systems.

In this section we describe the standardized speech teleservices that offer functionality for group communication: the Voice Broadcast Service (VBS) and the Voice Group Call

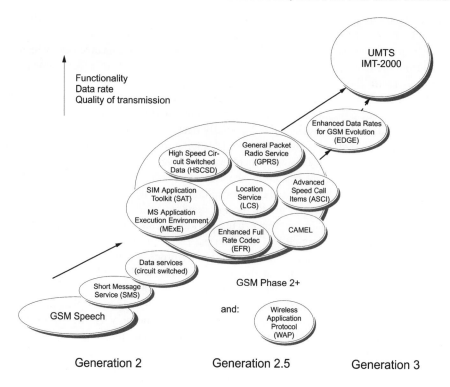

Figure 7.2 Evolution of GSM.

Service (VGCS). In addition, the Enhanced Multi-Level Precedence and Pre-emption Service (eMLPP) is used to assign and control priorities to users and their calls (e.g. for emergency calls). All of these services together are referred to as Advanced Speech Call Items (ASCI).

Voice broadcast service

VBS allows a user to broadcast a speech message to several other users within a certain geographical area. The user who initiates the call can only send ('speaker') and all others can only listen ('listeners').

Figure 7.3 gives a schematic illustration of a VBS scenario. Mobile users who are interested in a certain VBS group subscribe to it and will then receive broadcast calls of this group. Special permission is needed, however, for the right to send broadcast calls, i.e. for the right to act as a speaker. The subscribed VBS groups are stored on the user's SIM card, and if a subscriber does not want to receive VBS calls for a certain time, they can deactivate them. In addition to mobile GSM users, a predefined group of fixed telephone connections can also participate in the VBS service (e.g. dispatchers, supervisors, operators or recording machines).

System concept and group call register.   The area in which a speech broadcast call is offered is referred to as the group call area. As illustrated in Figure 7.4, in general, this area

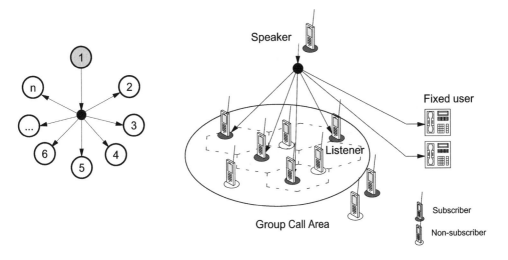

Figure 7.3 VBS scenario.

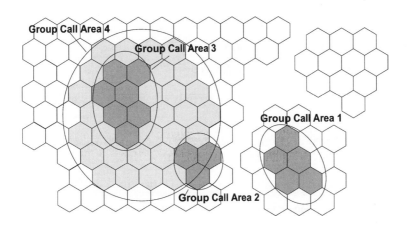

Figure 7.4 Some examples of group call areas.

consists of several cells. A group call area may comprise cells of several MSC areas and even of several PLMNs. One MSC is responsible for the handling of the VBS. It is called the anchor MSC. If a voice broadcast is also transmitted in cells that are not within the service area of this MSC (i.e. if the group call area also contains cells belonging to other MSCs), the MSCs of those cells are also involved. They are then denoted as relay MSCs.

The VBS-specific data are stored in a Group Call Register (GCR). Figure 7.5 shows the extended GSM system architecture. The GCR contains the broadcast call attributes for each VBS group, which are needed for call forwarding and authentication. Some examples follow.

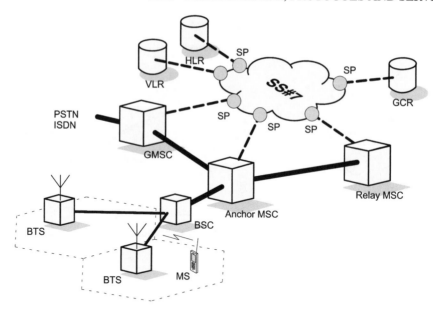

Figure 7.5 Extension of the GSM system architecture with the GCR.

- Which cells belong to the group call area?

- Which MSC is the responsible anchor MSC?

- In which cells are group members currently located, i.e. in which cells is a voice message to be broadcast?

- To which other MSCs is the voice message to be forwarded to reach all group members who are currently located in the group call area?

- To which external fixed telephone connections is the broadcast message addressed?

- Which fixed telephone connections are allowed to act as speakers?

Call establishment and logical channels.   A MS that intends to initiate a voice broadcast call sends a service request to the BSS. The request contains the group ID of the VBS group to be called. Thereupon, the responsible MSC queries the user's profile from the VLR and checks whether the user is allowed to act as speaker for the stated group. Next, some VBS-specific attributes are requested from the GCR. If the broadcast call should also be transmitted in cells that do not belong to the current MSC, an anchor MSC is determined. The anchor MSC then forwards the VBS attributes to all relay MSCs, which then request all affected BSCs to allocate a traffic channel in the respective cells, and to send out notification messages on the NCH (section 4.1). When a MS receives such a message and is also subscribing to the respective VBS group, it changes to the given traffic channel and listens to the voice broadcast in the downlink. The speaker is then informed about the successful connection setup and can

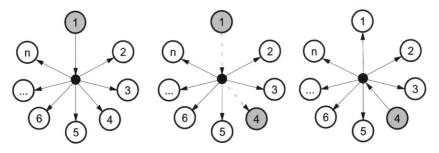

Figure 7.6 Group call scenario.

start talking. The notification message is periodically repeated on the NCH until the speaker terminates the call.

In contrast to the paging procedure in conventional GSM calls, the individual mobile users and their MSs are not explicitly addressed by an IMSI or TMSI but with the group ID of the VBS group. Furthermore, the MSs do not acknowledge the reception of VBS calls to the network. To realize the service, traffic channels are not allocated to individual subscribers, but the voice signal of the speaker is broadcast to all listening participants in a cell on one group channel. Thus, in each participating cell, only one full-rate channel is occupied (as in regular voice calls).

Voice Group Call Service

Another group communication service is the VGCS. The VGCS defines a closed user group communication service, where the right to talk can now be passed along within the group during a call by using a push-to-talk mechanism as in mobile radio. This principle is illustrated in Figure 7.6: user 1 initializes a group call and speaks, while the other users listen. Next, user 1 releases the channel and changes into listener mode. Now, each of the subscribers may apply for the right to become speaker. For example, user 4 requests the channel, and the network assigns it to them. They talk, release the channel and change back to listener mode. Finally, the group call is terminated by the initiator (in general). Whereas the information flow in the VBS is simplex, the VGCS can be regarded as a half-duplex system (compare Figures 7.3 and 7.6).

The fundamental concepts and entities of the VBS, e.g. the definition of group call areas, group IDs, the GCR and anchor and relay MSCs, are also used in the VGCS.

Logical channels. A traffic channel is allocated in each cell of the group call area that is involved in the VGCS. All group members listen to this channel in the downlink, and only the speaker uses it in the uplink. Therefore, in addition to the tasks for VBS calls, the network must also control uplink radio resources. The network indicates in the downlink to all MSs whether the uplink channel is in use or not. If the channel is free, the group members may send ABs. Collisions that occur with simultaneous requests are resolved, and the network chooses one user who obtains the channel and thus has the right to talk.

Table 7.2 Priority classes in eMLPP.

Class	Used by	Connection setup	Call interruption (pre-emption)	Example
A	Operator	Fast (1–2 s)	Yes	Highest priority; VBS/ VGCS emergency calls
B	Operator	Normal (<5 s)	Yes	Calls of operator
0	Subscriber	Normal (<5 s)	Yes	Emergency calls of users
1	Subscriber	Slow (<10 s)	Yes	
2	Subscriber	Slow (<10 s)	No	
3	Subscriber	Slow (< 10 s)	No	Standard priority
4	Subscriber	Slow (< 10 s)	No	Lowest priority

Enhanced Multi-Level Precedence and Pre-emption

Priority services enable a network to process calls with a priority class (precedence level). If the network load is high, calls with high priority can then be treated in a preferred manner and resources for low priority calls can be deallocated. In the extreme case, a call with low priority can be dropped because a call with high priority arrives (pre-emption).

The priorities in GSM are controlled by eMLPP. It is a supplementary service for point-to-point speech services as well as for VBS and VGCS. The principle of eMLPP is based on the Multi-Level Precedence and Pre-emption (MLPP) (ITU-T Recommendation Q.735, 2008) method used in SS#7. In doing so, MLPP has been enhanced with functions for priority control at the air interface. Table 7.2 lists all priority classes of eMLPP. In addition to the five precedence levels that are used in MLPP (classes 0–4), two additional levels with higher priority are defined (classes A and B). The table also shows whether a call with higher priority may terminate a call with lower priority. It is important to note that only the operator may use calls of class A and B, such that, for example, an emergency call over VBS or VGCS can be initiated in disaster situations. Calls of this class can only be employed within the service area of one MSC. The other five classes can be utilized within the entire PLMN and also in combination with the MLPP of ISDN. The highest priority call that a subscriber is allowed to use is stored on their SIM card and in the HLR.

7.4.2 New data services and higher data rates: HSCSD, GPRS and EDGE

As discussed in Chapter 8, the maximal data rate of 9.6 kbit/s for data services in conventional GSM is rather low compared with fixed networks. The desire for higher data rates in GSM networks was the motivation to introduce higher data rates and more flexible data services. Accordingly, one of the GSM standardization groups specified the HSCSD service. Later the new packet data service, GPRS, was standardized and has since been introduced by many network operators. It offers a genuine packet-switched bearer service at the air interface. As for HSCSD, new multislot-capable MSs are required (which can use, e.g., four time slots in the downlink and two time slots in the uplink). The GPRS chapter of this book (Chapter 8) discusses in detail the system architecture, protocols, air interface, multiple access and interworking with the Internet. Additional information can be found in Bettstetter *et al.* (1999), Brasche and Walke (1997), Faccin *et al.* (1999), Granbohm and Wiklund (1999), Kalden *et al.* (2000) and Walke (1999). Release 99 extended the GPRS standard with some new functions, e.g. point-to-multipoint services and prepaid services. Furthermore, existing functionality has been improved.

In order to increase the data rate and spectral efficiency on a single time slot, EDGE has been developed, as described in Chapter 8. It utilizes higher-order modulation schemes and a series of different coding schemes, in order to adapt the data rate perfectly to the current channel conditions of a terminal. Combining EDGE with GPRS and HSCSD, we obtain EGPRS and ECSD, respectively, now offering significantly higher data rates than pure HSCSD or GPRS.

7.5 Supplementary services in GSM Phase 2+

7.5.1 Supplementary services for speech

Most of the supplementary service characteristics known from ISDN have in one way or another already been implemented in GSM. The mobility of the users, however, creates the need for new supplementary services. Examples of supplementary services known from ISDN or newly defined are mobile access hunting, short message forwarding, multiple subscriber profile, call transfer or Completion of Calls to Busy Subscribers (CCBS).

The example of the CCBS service clearly shows how much the role of the HLR is changing from its original function as a database to a more active role as a service control component, similar to the Service Control Point (SCP) of the Intelligent Network (IN). The supplementary service CCBS basically realizes 'call back if busy'. If a called subscriber does momentarily not accept a call due to an ongoing connection, the calling subscriber can activate the supplementary service CCBS which causes the network to notify them at the end of the called subscriber's ongoing call and automatically set up the new connection. The subscriber mobility adds more complexity to the implementation of this service. In the fixed network, implementation would require the establishment of queues for call-back requests in the switching center of the calling and called subscribers, respectively. In a mobile network, this may involve additional switches, because after activation of the CCBS service, the calling subscriber may be roaming into another switching center area. If the implementation of the service were only in the MSC, either there could be a centralized solution or the queuing lists would have to be forwarded to the new MSC – which may even be in another network. The targeted solution is centralized in the HLR, which has to store the subscriber's callback queues (if existing) in addition to the current MSC designation. If the MS changes the MSC area, the callback queue is transferred to the new MSC. In this case, therefore, the HLR has to assume an additional server role and perform call control beyond the originally planned restriction to a pure database function.

7.5.2 Location service

GSM release 99 introduced a Location Service (LCS) making it possible to determine the exact location of a MS down to a few meters. One of the motivations for this service has been a law in the USA which demands the location of a person to be determined in the case of an emergency call.

From GSM mobility management, the network already knows the current cell of the user (cell identifier). However, this location accuracy is not sufficient in most cases, and therefore investigations have been made to find a more sophisticated solution. In the so-called Time Of Arrival (TOA) method, the network listens to handover ABs of the MS and is then able to triangulate its position. In contrast, using the Enhanced Observed Time Difference (E-OTD) method, the MSs now measure the time difference of received bursts from different base stations. Both methods only work if a MS has contact with at least three base stations. The accuracy of E-OTD schemes lies between 50 and 125 m, and that of TOA is worse (Buckingham, 1999). E-OTD schemes require a software update on the mobile equipment as well as modifications in the network, whereas for the TOA method it is mainly sufficient to modify network components. However, the functionality of TOA is provided by synchronization of the cellular network (using GPS or precise clocks at each BTS). This capability is currently not provided in asynchronous GSM networks. The most precise way to find the position of a MS is to integrate a GPS receiver into each piece of mobile equipment. The MS then receives its current position from GPS satellites. A substantial disadvantage of this approach is that MSs cannot always have intervisibility with GPS satellites (e.g. inside buildings). We observe that each of the three methods has its advantages and disadvantages.

In addition to the technical implementation of the location service, two new network nodes have been defined for this type of service: the Gateway Mobile Location Center (GMLC) and the Serving Mobile Location Center (SMLC). The GMLC acts as an interface to applications that use the positioning information of users in a specific way – so-called location-based or location-aware services. Examples are navigation services (such as 'Where is the closest gas station?') or virtual tourist guides ('What is the building on my left-hand side?'). A service provider stores, e.g., the locations of gas stations and sightseeing attractions in a database and adds other useful information. At the request of a mobile user, the provider can obtain the current position of the user from the GMLC and send back the requested information. Other location-based applications include location-based charging ('home zone'), vehicle tracking (e.g. stolen cars) and localized news, weather and traffic information.

7.6 Service platforms

The procedures for the development of the GSM standards required close cooperation between the involved manufacturers and network operators. The international standardization of services and interfaces led to a set of common successful performance characteristics in GSM networks, most prominently the international roaming capability. The more a performance criterion is standardized, the lower the costs of development and introduction are, since all manufacturers and operators contribute to paying the costs. On the other hand, the network operators desire service differentiation to be able to gain competitive advantages. The standardization of services and service performance criteria reduces the possibility for differentiation among competitors. Moreover, the time-to-market is often extended because of the prolonged process of standardization.

For these reasons, the service platform concept has been introduced in GSM on both the network and the terminal side. These platforms offer mechanisms, functions and protocols for the definition and control of services and applications. Those services/applications can be operator-specific, such that an international standardization process is not needed in general. The required generic functions can be made available in each MS and network node, and they can be used and combined in a flexible way for service execution.

The GSM supplementary services can be regarded as the simplest form of service platform usage. An extended concept are the so-called service nodes, such as a voice mail server and a SMS-SC. However, both concepts have significant disadvantages: supplementary services are subject to international standardization and, on the other hand, these services might not be available to roaming users in foreign networks, since network providers are not obligated to implement all supplementary services. The situation is similar with service nodes, which are often accessible only in the home network. We conclude that these two types of platforms allow the definition of vertical/operator-specific services in only a limited way, and their use in foreign networks is often not possible or rather complicated.

An extension of the platform concept, which has been taken up by ETSI in the Phase 2+ standardization, attempts to overcome this dilemma. Instead of specifying services and supplementary services directly or completely, only mechanisms are standardized which enable the introduction of new services. With this approach it is possible to restrict the implementation of a service to a few switches in the home network of a subscriber, whereas local (visited) switches have to provide only a fixed set of basic functions and the capability to communicate with the home network switch containing service logic.

This group of GSM standards within Phase 2+ is known under the name Support of Operator-Specific Services (SOSS), or also as Customized Applications for Mobile Network Enhanced Logic (CAMEL). The answer on the terminal side is the SIM Application Toolkit (SAT) and Mobile Station Execution Environment (MExE). They are explained in the following.

7.6.1 CAMEL: GSM and INs

Essentially, CAMEL represents a convergence of GSM and IN technologies. The fundamental concept of IN is to enable flexible implementation, the introduction and control of services in public networks and to use the idea of dividing the switching functionality into basic switching functionality, residing in Service Switching Points (SSPs) and centralized service control functionality, residing in Service Control Points (SCPs). Both network components communicate with each other over the signaling network using the generic SS#7 protocol extension called the Intelligent Network Application Part (INAP). This approach enables a centralized, flexible and rapid introduction of new services (Ambrosch *et al.*, 1989).

There are already some features in GSM which parallel an IN. Even though GSM standards use neither IN terminology nor IN protocols, i.e. INAP, the GSM network structure follows the IN philosophy (Laitinen and Rantale, 1995). In the GSM architecture, the separation into functional units such as MSC and HLR and the consistent use of SS#7 and its MAP extensions are in conformity with the IN architecture, which is split into SSPs and SCPs that communicate using INAP.

The philosophy of CAMEL is to proceed with the implementation of services in GSM in a similar way as in INs. This is reflected in separating a set of basic call processing functions in the MSC or GMSC (which act as SSPs), from the intelligent service control functions (SCP) in the home network of the respective subscriber. The HLR in a GSM network already has functions similar to the SCP, especially with regards to supplementary services. Beyond that, the CAMEL approach provides its own dedicated SCPs. Imagine specialized SCPs for the translation of abbreviated numbers in Virtual Private Networks (VPNs) or for future extended SMSs. With this configuration, the service implementation with its service logic is needed only once, namely in the home network SCP. The network operator offering the service thus has sole control over the features and performance range of the service. Owing to the complete range of generic functions that have to be provided at each SSP (MSC, GMSC, etc.), new services can immediately be provided in each network, and an uninterrupted service availability is guaranteed for roaming subscribers. The sole responsibility and control for the introduction of new services lies in SOSS/CAMEL with the operator of the home network, the contract partner of the subscriber. This opens new competitive possibilities among network operators. Operator-specific services can be introduced rapidly without having to go through the standardization process, and yet they are available worldwide.

Figure 7.7 shows the resulting architecture. The CAMEL specification requires a GSM-specific version of IN. Similar to the IN approach, GSM defines a basic call processing function as GSM Service Switching Function (gsmSSF) and a service logic function GSM Service Control Function (gsmSCF). In addition to the MAP signaling interfaces already existing in Phase 1 and Phase 2 for communication between visited and home networks (GMSC, VLR, HLR), new signaling interfaces are needed for communication between basic switching and service logic in the visited and home networks. For this signaling, a new application part of SS#7 is being specified, the CAMEL Application Part (CAP), which assumes the functions similar to INAP. These functions and protocols represent the basic structure for the realization of intelligent services and their flexible introduction.

The prerequisites for CAMEL are the definition of a standardized extended call model with appropriate trigger points, and the specification of the generic range of services which must be provided by the SSP. More precisely, the new extended call model must also include a model of subscriber behavior, because in addition to normal call processing aspects, it also contains events such as location updating. For each subscriber, this behavior model is stored in the HLR and supplied from the home network to the currently visited SSP/MSC. In this way, the SSP/MSC has a set of trigger points with corresponding SCP addresses for each subscriber roaming in its area. When a trigger condition is satisfied, the call and transaction processing in the SSP/MSC is interrupted and the SCP is notified. The SCP can now analyze the context and, according to the service implementation, give instructions to the SSP to perform particular functions. Typical functions the SSP has to implement are call forwarding, call termination or other stimuli to the subscriber (Mouly and Pautet, 1995). Based on this behavior model and the corresponding control protocol between mobile SSP and home SCP, which are connected through a set of generic SSP functions, we can expect to see a large variety of operator-specific services in the future.

7.6.2   Service platforms on the terminal side

SIM application toolkit

SAT has been a further step toward provider-specific vertical services. The GSM SIM card is provided completely by the network operator, in particular because it contains security functions. From this fact, the basic approach arose to equip the SIM card with additional, operator-specific functions. Without a standardized interface to the mobile equipment,

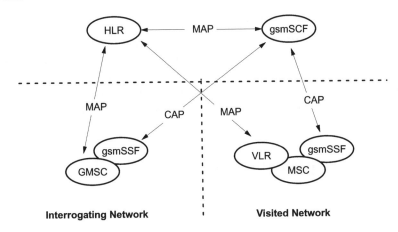

Figure 7.7 Functional architecture for CAMEL.

this was only possible in a very limited manner and only in close cooperation with equipment manufactures. SAT removes these restrictions by defining a standardized interface between mobile equipment and SIM card. In this way, operator-specific applications can run on the SIM card and can thereby control clearly defined, selected functions of the terminal. Corresponding applications can be carried out in the PLMN or even outside the PLMN on dedicated servers making it possible to implement completely new services. The communication between the SIM card application and its counterpart in the network is currently implemented over SMS, but in the near future other bearer services (in particular, GPRS) are also possible. The functions defined in the SAT framework can be categorized into SIM data download and proactive SIM. The functional interface between SIM card and terminal is through proactive SIM mechanisms. They include:

- display of text;

- transmission of SMS messages;

- connection setup (speech and data) triggered by the SIM card;

- playing of sounds in the mobile equipment;

- read-out of local information from the equipment into the SIM card.

With these mechanisms, a broad variety of new features can be offered, for example, download of data to the SIM card. This includes the download of new or existing commands and applications to be installed. With the toolkit, the SIM card is able to display new, operator-specific menu options to the user, and to read out user actions from the mobile equipment. Most far-reaching are the functions for call control, where each number typed in can be analyzed by the SIM card. This allows for operator-specific treatment of telephone numbers, e.g., the mapping of numbers or barring functions. In a further standardization step, the functions of SAT have been enhanced with security and encryption mechanisms. SAT-capable MSs have been available for a few years.

Mobile station application execution environment

Of similar scope is the MExE, which implements a generic application platform in the terminal. The most important components are a virtual machine for execution of Java code and WAP. Both techniques open the door for a variety of new services and applications. With a virtual machine running on the mobile terminal, applications can be uploaded and executed. This demands a high computational effort in the MSs. The WAP is explained in the following section.

7.7   Wireless application protocol

WAP is a major step in building the wireless Internet, where people on-the-go can access the Internet through their wireless devices to get information such as emails, news headlines, stock reports, map directions and sports scores when they need it and where they need.

(Chuck Parish, Founding Member and Chairman (1998–1999) WAP Forum)

WAP is regarded as an important step in the move of today's GSM networks towards a 'mobile Internet'. During the last few years, WAP has been developed and standardized by the WAP Forum 1999a; 1999b. This industry consortium was founded by Nokia, Ericsson; Phone.com (formerly Unwired Planet) and Motorola in December 1997 and has several hundred members today.

The philosophy of WAP is to transfer Internet content and other interactive services to MSs to make them accessible to mobile users. For this purpose, WAP defines a system architecture, a protocol family, and an application environment for the transmission and display of WWW-like pages for mobile devices.

The motivation for the development of WAP was the fundamental restrictions posed by mobile equipment and cellular networks in comparison to PCs and fixed wired networks. These are, in particular, the limited opportunities for display and input (small displays, number keypad and no mouse) as well as the limited memory and processing power. Furthermore, the power consumption of a MS should be as low as possible. On the network side, it is clear that the wireless transmission has lower bandwidth, a higher bit error probability and less stable connections than wired networks.

The protocols and the application environment defined for WAP consider these limitations. The protocols of the WAP architecture are basically a modification, optimization and enhancement of the IP stack used in the World Wide Web for use in mobile and wireless environments. WAP focuses on applications tailored to the capabilities of cellular phones and the needs of mobile users. One can say that WAP 'creates an information Web for cellular phones, distinct from the PC-centric Web' (Goodman, 2000).

7.7.1   Wireless markup language

With respect to the mentioned requirements, the Wireless Markup Language (WML) has been developed. It represents a pendant to the Hypertext Markup Language (HTML) used in the World Wide Web. WML is defined as a document type of the meta language the Extensible Markup Language (XML). It contains some phone-specific tags and requires only a phone

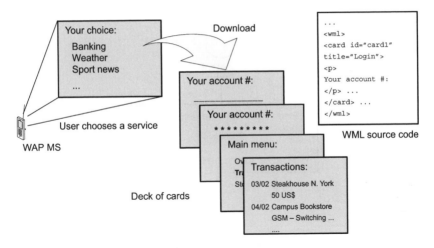

Figure 7.8 WAP example.

keypad for input. For display of monochrome graphics the Wireless Bitmap (WBMP) format has been defined.

A microbrowser, which is running on each WAP terminal, interprets the received WML documents and displays their content (text, pictures, links) to the user. Such a microbrowser is also referred to as a WML browser and is the pendant to a Web browser used in PCs.

In doing so, the presentation of WML documents is not limited to classical mobile telephones, but there also exist WML browsers for other devices, such as for PDAs under the operating systems PalmOS, Windows CE or EPOC systems. These devices may be linked over infrared or Bluetooth (Bluetooth SIG, 2008) with a GSM MS, or they have their own GSM/GPRS air interface.

WML documents are organized into cards and decks (see Figure 7.8). When a subscriber chooses a service, a deck of cards is downloaded to the MS. The user can then view these cards with their WML browser, make inputs and navigate between the cards. Each card is designed for one user interaction.

Figure 7.9 illustrates how an automatic parallel creation of WML and HTML documents may look (WAP Forum, 1999c). The World Wide Web Consortium (W3C) currently specifies the Extensible Style Language (XSL) (WWW Consortium, 2008). Using XSL style sheets, WML and HTML documents can be automatically generated from content written in XML.

7.7.2 Protocol architecture

The WAP architecture is shown in Figure 7.10. As mentioned before, WAP is based on the WWW protocol stack and adjusts those protocols to the requirements of wireless transmission and small portable devices.

For applications, a uniform microbrowser environment has been specified: the Wireless Application Environment (WAE). It comprises the following functionality and formats:

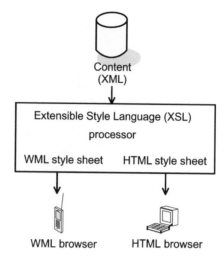

Figure 7.9 Generation of WML and HTML documents.

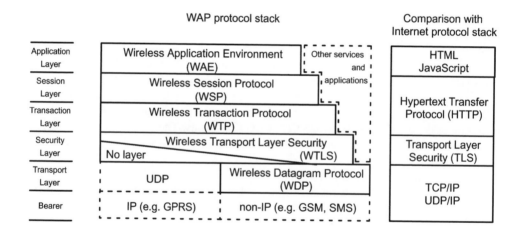

Figure 7.10 WAP architecture.

- the WML;

- a simple script language WMLScript, which is based on JavaScript;

- programming interfaces for control of telephony services (Wireless Telephony Application (WTA) interface); and

- data formats for pictures, electronic business cards (vCard) and entries of the phone directory and calendar (vCalendar).

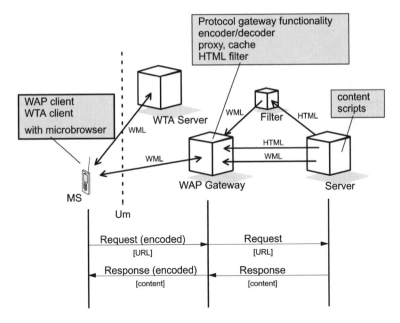

Figure 7.11 WAP system architecture and REQUEST/RESPONSE transaction.

The WTA interface allows the microbrowser to interact with telephony functions. For example, it specifies how calls are initiated from the microbrowser or how entries from the phone directory are sent out.

The main task of the Wireless Session Protocol (WSP) is the establishment and termination of a session between the MS and the WAP gateway (Figure 7.11). Thereby, connection-oriented (over WTP) as well as connectionless (over datagram services, e.g. WTP) sessions are defined. In the case of a radio connection breaking down, the session can be stopped for a certain period of time and resumed later.

The Wireless Transaction Protocol (WTP) is a lightweight transaction-oriented protocol. Its task is to guarantee the reliable exchange of the MS's request and the WAP gateway's response messages (Figure 7.11). It thus constitutes the basis for interactive browsing. WTP includes functions for the acknowledgement of messages, retransmission of erroneous or lost messages, and the removal of duplicate messages. In addition, an acknowledged and an unacknowledged datagram service is defined for push services, where the server can send content to a MS without an initiating request from the mobile user. The server may send an emergency warning, for example.

Optionally, the Wireless Transport Layer Security (WTLS) protocol may be employed. It is based on the Transport Layer Security (TLS) protocol, which is used in the Internet and was formerly known as Secure Socket Layer (SSL). WTLS offers basic security functions, such as data integrity, encryption, user identity confidentiality and authentication between the server and MS. Moreover, protection against denial-of-service attacks is provided. The functionality of WTLS can be made effective (or not) according to the application and security of the used network. For example, if an application already uses strict security techniques, the complete

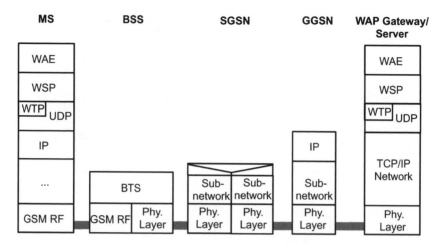

Figure 7.12 Protocol architecture WAP over GPRS (compare with Figure 8.7).

scope of WTLS functions will not be needed. It is worth mentioning that WTLS can also be used for secure data transfer between two MSs (e.g. for authenticated exchange of electronic business cards).

The WAP transport protocol is known as the Wireless Datagram Protocol (WDP). It is for example, used instead of UDP for bearer services that are not based on IP (Figure 7.10). GSM bearer services for WAP can be either circuit-switched data services (e.g. SMS) or the GPRS, which of course offers faster data transfer and volume-based billing.

Figure 7.10 indicates that nonWAP protocols can also access specific layers of the WAP stack. Furthermore, not all WAP protocols must be used always. Certain applications may, for example, only require the services of WTP and underlying layers. Figure 7.12 gives an example, where WAP comes into operation over GPRS as a bearer service.

7.7.3   System architecture

Figure 7.11 gives a schematic illustration of a typical WAP system architecture. The principle of how content is stored in a distributed way within the network and finally offered to the user is similar to the principle of the WWW. Servers store content directly as a WML document or content is generated with scripts. MSs download this content from the server to their microbrowser, which then presents them to the user. In theory, it is possible to store content in HTML and subsequently convert it to WML, however, in practice, applications and content offered directly in WML are much better suited (WAP Forum, 1999a).

As shown in Figure 7.11, a WAP Gateway acts as an interface between external servers and the MSs. Its main tasks include:

- conversion of requests from the WAP protocol stack to the WWW protocol stack (HTTP over TCP/IP) and vice versa (i.e. a protocol gateway functionality);

- encoding and decoding of WML documents into a binary. format

The WAP gateway also represents a proxy server and acts as a cache for frequently requested contents.

The following example illustrates the transaction procedure between a mobile user, the WAP gateway and an external server: A subscriber intends to view a document which is offered on a server. Their WML browser sends a WSP REQUEST to the appropriate address of the server. The request is forwarded to the WAP gateway, which then converts it into an HTTP REQUEST and contacts the server. Next, the server transmits the requested content in WML format to the WAP gateway, which writes the content into its cache and sends it in binary-encoded form to the MS. The latter presents the first card of the deck on the microbrowser to the user. If the external server transmitted the document in HTML format, the gateway would convert it into WML format.

7.7.4 Services and applications

The first specification of WAP was released by the WAP Forum in April 1998. Version 1.1 followed in June 1999 and Version 1.2 in December 1999. WAP terminals have been introduced around February 1999 for the first time, and today there exist a broad variety of WAP products: mobile equipment, gateways, development tools, WML browsers and editors.

In addition to the technical implementation in the network and the development of new WAP-capable mobile equipment, innovative WAP services are, in particular, in demand. These days, several information services are offered over WAP. Subscribers can retrieve news, weather forecasts, stock reports and local restaurant and event guides with their WAP phone. Furthermore, mobile e-commerce services (e.g. ticket reservation, mobile banking and online auctions) are becoming increasingly popular. There is much scope left for new applications. Push services, for example, may transmit important information to MSs without the need to request them actively. Highly interesting are so-called location-based services, in which the service knows the current physical location of the user and may use this information in a specific way. Navigation services with displayed maps on the browser or virtual tourist guides ('I would like to have information about the building on my left-hand side') are two examples.

8

Improved data services in GSM: GPRS, HSCSD and EDGE

The main data service used in modern GSM systems is the GPRS. Therefore, the major part of this chapter is devoted to GPRS. However, HSCSD is also of importance. HSCSD was available earlier than GPRS and was therefore chosen by some network providers in order to provide higher data rate services in GSM as fast as possible and it is still used in many networks. A third important aspect of providing improved data rates in GSM systems is EDGE, which helps to increase the data rate of both GPRS and HSCSD by allowing for higher-order modulation schemes when the signal strength is sufficiently high. Let us now first focus on GPRS in the following.

8.1 GPRS

Packet data transmission has already been standardized in GSM phase 2, offering access to the Packet Switched Public Data Network (PSPDN); see Appendix A. However, on the air interface such access occupies a complete circuit switched traffic channel for the entire call period. In the case of bursty traffic (e.g. Internet traffic), such access leads to a highly inefficient resource utilization. It is obvious that in this case, packet switched bearer services result in a much better utilization of the traffic channels. This is because a packet channel will only be allocated when needed and will be released after the transmission of the packets. With this principle, multiple users can share one physical channel (statistical multiplexing).

In order to address these inefficiencies, GPRS has been developed in GSM phase 2+. It offers a genuine packet-switched bearer service for GSM also at the air interface. GPRS is thus a huge improvement and simplification of the wireless access to packet data networks. Networks based on IP (e.g. the global Internet or private/corporate intranets) and X.25 networks are supported. In order to introduce GPRS to existing GSM networks, several modifications and enhancements must be made in the network infrastructure as well as in the MSs.

GSM – Architecture, Protocols and Services Third Edition   J. Eberspächer, H.-J. Vögel, C. Bettstetter and C. Hartmann
© 2009 John Wiley & Sons, Ltd

Users of GPRS benefit from higher data rates and shorter access times. In conventional GSM, the connection setup takes several seconds and rates for data transmission are restricted to 9.6 kbit/s. GPRS, in practice, offers almost ISDN-like data rates up to approximately 40–50 kbit/s and session establishment times below one second. Furthermore, GPRS supports a more user-friendly billing than that offered by circuit-switched data services. In circuit-switched services, billing is based on the duration of the connection. This is unsuitable for applications with bursty traffic, since the user must pay for the entire airtime even for idle periods when no packets are sent (e.g. when the user reads a Web page). In contrast to this, with packet-switched services, billing can be based on the amount of transmitted data (e.g. Mbyte) and the Quality of Service (QoS). The advantage for the user is that they can be 'online' over a long period of time but will be billed mainly based on the transmitted data volume. The network operators can utilize their radio resources in a more efficient way and simplify the access to external data networks.

The structure of this chapter is as follows.[1] Section 8.1.1 gives an overview of the GPRS system architecture and explains the fundamental functionality. Next, in section 8.1.2, we describe the offered services and the QoS parameters. Section 8.1.3 explains the session and mobility management and routing. It answers for example, the following questions. How does a GPRS MS register with the network? How does the network keep track of the MS's location? Section 8.1.4 gives an overview of the GPRS protocol architecture and briefly introduces the protocols developed for GPRS. Next, an example of a GPRS–Internet interconnection is given (section 8.1.6). Section 8.1.7 discusses the air interface, including the multiple access concept and radio resource management. Moreover, the logical channels and their mapping onto physical channels are explained. Section 8.1.7 considers GPRS channel coding. GPRS security issues are treated in section 8.1.8 and, finally, a brief summary of the main features of GPRS is given.

8.1.1 System architecture of GPRS

In order to integrate GPRS into the existing GSM architecture (Chapter 3), a new class of network nodes, called GPRS Support Nodes (GSNs), has been introduced. GSNs are responsible for the delivery and routing of data packets between the MSs and external Packet Data Networks (PDNs). Figure 8.1 illustrates the resulting system architecture.

A Serving GPRS Support Node (SGSN) delivers data packets from and to the MSs within its service area. Its tasks include packet routing and transfer, functions for attach/detach of MSs and their authentication, and logical link management. The location register of the SGSN stores location information (e.g. current cell, current VLR) and user profiles (e.g. IMSI, address used in the packet data network) of all GPRS users registered with this SGSN.

A Gateway GPRS Support Node (GGSN) acts as an interface to external packet data networks (e.g. to the Internet). It converts GPRS packets coming from the SGSN into the appropriate Packet Data Protocol (PDP) format (i.e. IP or X.25) and sends them out on the corresponding external network. In the other direction, the PDP address of incoming data packets (e.g. the IP destination address) is converted into the GSM address of the destination

---

[1]Parts of this chapter are based on: Ch. Bettstetter, H.-J. Vögel, J. Eberspächer. GSM Phase 2+ General Packet Radio Service GPRS: Architecture, Protocols, and Air Interface. *IEEE Communications Surveys*, Special Issue on Packet Radio Networks, vol. 2, no. 3, 1999, which can be obtained at http://www.comsoc.org/pubs/surveys. © 1999 IEEE.

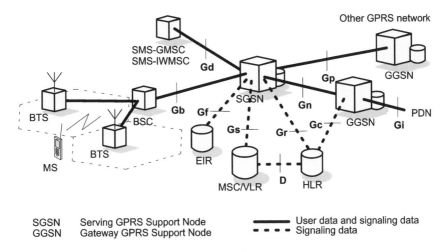

SGSN   Serving GPRS Support Node   ▬▬▬ User data and signaling data
GGSN   Gateway GPRS Support Node   ▬ ▬ ▬ Signaling data

Figure 8.1 GPRS system architecture and interfaces.

user. The readdressed packets are sent to the responsible SGSN. For this purpose, the GGSN stores the current SGSN addresses and profiles of registered users in its location register.

In general, there is a many-to-many relationship between the SGSNs and the GGSNs: a GGSN is the interface to an external network for several SGSNs; an SGSN may route its packets to different GGSNs.

Figure 8.1 also shows the interfaces between the GPRS support nodes and the GSM network. The Gb interface connects the BSC with the SGSN. Via the Gn and the Gp interfaces, user and signaling data are transmitted between the GSNs. The Gn interface is used if SGSN and GGSN are located in the same PLMN, whereas the Gp interface is used if they are in different PLMNs.

All GSNs are connected via an IP-based GPRS backbone network. Within this backbone, the GSNs encapsulate the PDN packets and transmit (tunnel) them using the so-called GPRS Tunneling Protocol (GTP). In principle, we can distinguish between two kinds of GPRS backbones.

- Intra-PLMN backbones are IP-based networks owned by the GPRS network provider connecting the GSNs of the GPRS network.

- Inter-PLMN backbone networks connect GSNs of different GPRS networks. They are installed if there is a roaming agreement between two GPRS network providers.

Figure 8.2 shows, how two intra-PLMN backbone networks of different PLMNs are connected with an inter-PLMN backbone. The gateways between the PLMNs and the external inter-PLMN backbone are called Border Gateways (BGs). Their main task is to perform security functions in order to protect the private intra-PLMN backbones against unauthorized users and attacks. The illustrated routing example is explained later.

The Gn and Gp interfaces are also defined between two SGSNs. This allows the SGSNs to exchange user profiles when a MS moves from one SGSN area to another.

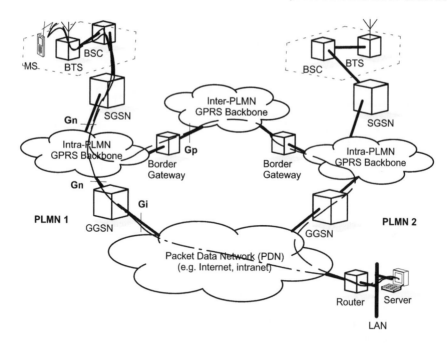

Figure 8.2 GPRS system architecture, interfaces and routing example.

Across the Gf interface, the SGSN may query and check the IMEI of a MS trying to register with the network.

The Gi interface connects the PLMN with external PDNs. In the GPRS standard, interfaces to IP (IPv4 and IPv6) and X.25 networks are supported.

GPRS also adds some more entries to the GSM registers. For MM, the user's entry in the HLR is extended with a link to its current SGSN. Moreover, their GPRS-specific profile and current PDP address(es) are stored. The Gr interface is used to exchange this information between HLR and SGSN. For example, the SGSN informs the HLR about the current location of the MS. When a MS registers with a new SGSN, the HLR will send the user profile to the new SGSN. In a similar manner, the signaling path between GGSN and HLR (Gc interface) may be used by the GGSN to query the location and profile of a user who is unknown to the GGSN.

In addition, the MSC/VLR may be extended with functions and register entries which allow efficient coordination between packet-switched (GPRS) and conventional circuit-switched GSM services. Examples for this are combined GPRS and GSM location updates and combined attachment procedures. Moreover, paging requests of circuit-switched GSM calls can be performed via the SGSN. For this purpose, the Gs interface connects the registers of SGSN and MSC/VLR.

Finally, it is worth mentioning that it is possible to exchange messages of the SMS via GPRS. The Gd interface interconnects SMS-GMSC with the SGSN.

8.1.2 Services

Bearer services and supplementary services

The bearer services of GPRS offer end-to-end packet switched data transfer to mobile subscribers. Currently, a Point-to-Point (PTP) service is specified, which comes in two variants: a connectionless mode (PTP Connectionless Network Service (PTP-CLNS), e.g. for IP) and a connection-oriented mode (PTP Connection Oriented Network Service (PTP-CONS), e.g. for X.25).

It is possible to use IP multicast routing protocols (see, e.g., Sahasrabuddhe and Mukherjee (2000)) over GPRS. Packets addressed to an IP multicast group will then be routed to all group members. Furthermore, SMS messages can be sent and received over GPRS.

Based on these standardized services, GPRS providers may offer additional nonstandardized services. Examples are access to information databases, messaging services (via store-and-forward mailboxes) and transaction services (e.g. credit card validations and electronic monitoring/surveillance systems). The most important application scenario, however, is the wireless access to the World Wide Web and to corporate intranets as well as e-mail communication.

Quality of service

The QoS requirements for the variety of mobile data applications, in which GPRS is used as transmission technology, are very diverse (for example, compare the requirements of real-time video conferencing with those of e-mail transfer with respect to packet delay and error-free transmission). Support of different QoS classes is therefore an important feature to support a broad variety of applications but still preserve radio and network resources in an efficient way. Moreover, QoS classes enable providers to offer different billing options. The billing can be based on the amount of transmitted data, the service type itself and the QoS profile. At the moment, four QoS parameters are defined in GPRS: service precedence, reliability, delay and throughput. Using these parameters, QoS profiles can be negotiated between the mobile user and the network for each session, depending on the QoS demand and the currently available resources.

The service precedence is the priority of a service (in relation to other services). There exist three levels of priority: high, normal and low. In the case of a heavy traffic load, for example, packets of low priority will be discarded first.

The reliability indicates the transmission characteristics required by an application. Three reliability classes are defined (see Table 8.1), which guarantee certain maximum values for the probability of packet loss, packet duplication, mis-sequencing and packet corruption (i.e. undetected error in a packet).

The delay parameters define maximum values for the mean delay and the 95th percentile delay (see Table 8.2). The latter is the maximum delay guaranteed in 95% of all transfers. Here, 'delay' is defined as the end-to-end transfer time between two communicating MSs or between a MS and the Gi interface to an external network, respectively. This includes all delays within the GPRS network, e.g., the delay for request and assignment of radio resources, transmission over the air interface and the transit delay in the GPRS backbone network. Delays outside the GPRS network, e.g. in external transit networks, are not taken

Table 8.1 Probability of various outcomes with the three reliability classes.

Class	Lost packet	Duplicated packet	Out of sequence packet	Corrupted packet
1	10^{-9}	10^{-9}	10^{-9}	10^{-9}
2	10^{-4}	10^{-5}	10^{-5}	10^{-6}
3	10^{-2}	10^{-5}	10^{-5}	10^{-2}

Table 8.2 Delay classes.

Class	128 byte packet		1024 byte packet	
	Mean delay (s)	95% delay (s)	Mean delay (s)	95% delay (s)
1	<0.5	<1.5	<2	<7
2	<5	<25	<15	<75
3	<50	<250	<75	<375
4	Best effort	Best effort	Best effort	Best effort

into account. Table 8.2 lists the four defined delay classes and their parameters for a 128 byte and 1024 byte packet, respectively.

Finally, the throughput parameter specifies the maximum/peak bit rate and the mean bit rate.

Simultaneous use of packet-switched and circuit-switched services

In a GSM/GPRS network, conventional circuit-switched services (GSM speech, data and SMS) and GPRS services can be used in parallel. The GPRS standard defines three classes of MSs: MSs of class A fully support simultaneous operation of GPRS and conventional GSM services. Class B MSs are able to register with the network for both GPRS and conventional GSM services simultaneously and listen to both types of signaling messages, but can only use one of the service types at a given time. Finally, class C MSs can attach for either GPRS or conventional GSM services at a given time. Simultaneous registration (and use) is not possible, except for SMS messages, which can be received and sent at any time.

8.1.3   Session management, mobility management and routing

In this section we describe how a MS registers with the GPRS network and becomes known to an external packet data network. We show how packets are routed to or from MSs, and how the network keeps track of the user's current location.

Attachment and detachment procedure

Before a MS can use GPRS services, it must attach to the network (similar to the IMSI attach used for circuit-switched GSM services). The MS's ATTACH REQUEST message is sent to

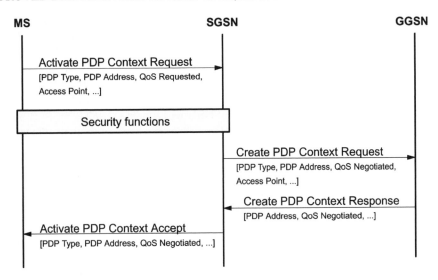

Figure 8.3 PDP context activation.

the SGSN. The network then checks whether the user is authorized, copies the user profile from the HLR to the SGSN, and assigns a Packet Temporary Mobile Subscriber Identity (P-TMSI) to the user. This procedure is called GPRS attach. For MSs using both circuit-switched and packet-switched services, it is possible to perform combined GPRS/IMSI attach procedures. The disconnection from the GPRS network is called GPRS detach. It can be initiated by the MS or by the network.

Session management and PDP context

To exchange data packets with external PDNs after a successful GPRS attach, a MS must apply for an address used in the PDN. In general, this address is called a PDP address. In case the PDN is an IP network, this will be an IP address.

For each session, a so-called PDP context is created, which describes the characteristics of the session. It contains the PDP type (e.g. IPv4), the PDP address assigned to the MS (e.g. an IP address), the requested QoS class and the address of a GGSN that serves as the access point to the external network. This context is stored in the MS, the SGSN and the GGSN. Once a MS has an active PDP context, it is 'visible' to the external network and can send and receive data packets. The mapping between the two addresses (PDP ↔ GSM address) makes the transfer of data packets between MS and GGSN possible.

The allocation of a PDP address can be static or dynamic. In the first case, the MS permanently owns a PDP address, which has been assigned by the network operator of the user's home PLMN. Using a dynamic addressing concept, a PDP address is assigned upon activation of a PDP context, i.e. each time a MS attaches to the network it will in general get a new PDP address, and after its GPRS detach this PDP address will again be available to other MSs. The PDP address can be assigned by the user's home-PLMN operator (dynamic home-PLMN PDP address) or by the operator of the visited network (dynamic visited-PLMN PDP address). The GGSN is responsible for the allocation and deactivation of the addresses.

Figure 8.3 shows the PDP context activation procedure initialized by the MS. Using the message `ACTIVATE PDP CONTEXT REQUEST`, the MS informs the SGSN about the requested PDP context. If a dynamic address is requested, the parameter `PDP ADDRESS` will be left empty. Afterwards, the usual GSM security functions (e.g. authentication of the user) are performed. If access is granted, the SGSN will send a `CREATE PDP CONTEXT REQUEST` to the affected GGSN. The GGSN creates a new entry in its PDP context table, which enables the GGSN to route data packets between the SGSN and the external PDN. It confirms this to the SGSN with a message `CREATE PDP CONTEXT RESPONSE`, which also contains the dynamic PDP address (if needed). Finally, the SGSN updates its PDP context table and confirms the activation of the new PDP context to the MS (`ACTIVATE PDP CONTEXT ACCEPT`).

It is also worth mentioning that the GPRS standard supports anonymous PDP context activation, which is useful for special applications such as prepaid services. In such a session, the user (i.e. the IMSI) using the PDP context remains unknown to the network. Security functions as shown in Figure 8.3 are skipped. Only dynamic address allocation is possible in this case.

Routing

In Figure 8.2 we give an example of how packets can be routed in GPRS. We assume that the packet data network is an IP network.

A GPRS MS located in PLMN1 sends IP packets to a Web server connected to the Internet. The SGSN which the MS is registered with encapsulates the IP packets coming from the MS, examines the PDP context, and routes them through the GPRS backbone to the appropriate GGSN. The GGSN decapsulates the IP packets and sends them out on the IP network, where IP routing mechanisms transfer the packets to the access router of the destination network. The latter delivers the IP packets to the host.

Let us assume that the MS's home-PLMN is PLMN2 and that its IP address has been assigned from the PLMN2 address space – either in a dynamic or static way. When the Web server now addresses IP packets to the MS, they are routed to the GGSN of PLMN2 (the home-GGSN of the MS). This is because the MS's IP address has the same network prefix as the IP address of its home-GGSN. The GGSN queries the HLR and obtains the information that the MS is currently located in PLMN1. In the following, it encapsulates the incoming IP packets and tunnels them through the inter-PLMN GPRS backbone to the appropriate SGSN in PLMN1. The SGSN decapsulates the packets and delivers them to the MS.

Location management

As in circuit-switched GSM, the main task of location management is to keep track of the user's current location, so that incoming packets can be routed to their MS. For this purpose, the MS frequently sends location update messages to its SGSN.

How often should a MS send such a message? If it updates its current location (e.g. its cell) rather seldom, the network must perform a paging process in order to search the MS when packets are coming in. This will result in a significant delivery delay. On the other hand, if location updates happen very often, the MS's location is well known to the network

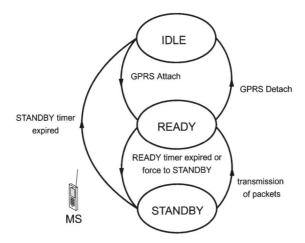

Figure 8.4 State model of a GPRS MS.

(and thus the packets can be delivered without any additional paging delay), but quite a lot of uplink radio bandwidth and battery power is used for MM in this case. Thus, a good location management strategy must be a compromise between these two extreme methods.

For this reason, a state model for GPRS MSs has been defined (shown in Figure 8.4). In IDLE state the MS is not reachable. Performing a GPRS attach, it turns into READY state. With a GPRS detach it may deregister from the network and fall back to IDLE state, and all PDP contexts will be deleted. The STANDBY state will be reached when a MS does not send any packets for a long period of time, and therefore the READY timer (which was started at GPRS attach and is reset for each incoming and outgoing transmission) expires. The location update frequency depends on the state in which the MS currently is. In IDLE state, no location updating is performed, i.e. the current location of the MS is unknown. If a MS is in READY state, it will inform its SGSN of every movement to a new cell. For the location management of a MS in STANDBY state, a GSM LA is divided into so-called Routing Areas (RAs). In general, an RA consists of several cells. The SGSN will only be informed, when a MS moves to a new RA; cell changes will not be indicated.

To find out the current cell of a MS that is in STANDBY state, paging of the MS within a certain RA must be performed (section 8.1.7). For MSs in READY state, no paging is necessary.

Whenever a MS moves to a new RA, it sends a `ROUTING AREA UPDATE REQUEST` to its assigned SGSN (Figure 8.5). The message contains the Routing Area Identity (RAI) of its old RA. The BSS adds the CI of the new cell to the request, from which the SGSN can derive the new RAI. Two different scenarios are possible:

- intra-SGSN routing area update (Figure 8.5);

- inter-SGSN routing area update (Figure 8.6).

In the Intra-SGSN case, the MS has moved to an RA which is assigned to the same SGSN as the old RA. In this case, the SGSN has already stored the necessary user profile and

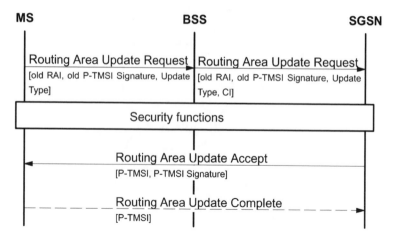

Figure 8.5 Intra-SGSN routing area update.

can immediately assign a new P-TMSI (ROUTING AREA UPDATE ACCEPT). Since the routing context does not change, there is no need to inform other network elements, such as GGSN or HLR.

In the inter-SGSN case, the new RA is administered by a different SGSN than the old RA. The new SGSN realizes that the MS has entered its area and requests the old SGSN to send the PDP contexts of the user (SGSN CONTEXT REQUEST, SGSN CONTEXT RESPONSE, SGSN CONTEXT ACKNOWLEDGE). Afterward, the new SGSN informs the involved GGSNs about the user's new routing context (UPDATE PDP CON-TEXT REQUEST, UPDATE PDP CONTEXT RESPONSE). In addition, the HLR and (if needed) the MSC/VLR are informed about the user's new SGSN number (UPDATE LOCA-TION,..., UPDATE LOCATION ACKNOWLEDGE; LOCATION UPDATE REQUEST, LOCATION UPDATE ACCEPT).

In addition to pure RA updates, there also exist combined RA/LA updates. They are performed whenever a MS using GPRS as well as conventional GSM services moves to a new LA. The MS sends a ROUTING AREA UPDATE REQUEST to the SGSN and uses a parameter update type to indicate that an LA update is needed. The message is then forwarded from the SGSN to the VLR.

To sum up, we can say that GPRS mobility management consists – as with GSM mobility management – of two levels: micro mobility management tracks the current RA or cell of the user; macro mobility management keeps track of the user's current SGSN and stores it in the HLR, VLR and GGSN.

8.1.4   Protocol architecture

Transmission plane

Figure 8.7 illustrates the protocol architecture of the GPRS transmission plane. The protocols offer transmission of user data and its associated signaling (e.g. for flow control, error

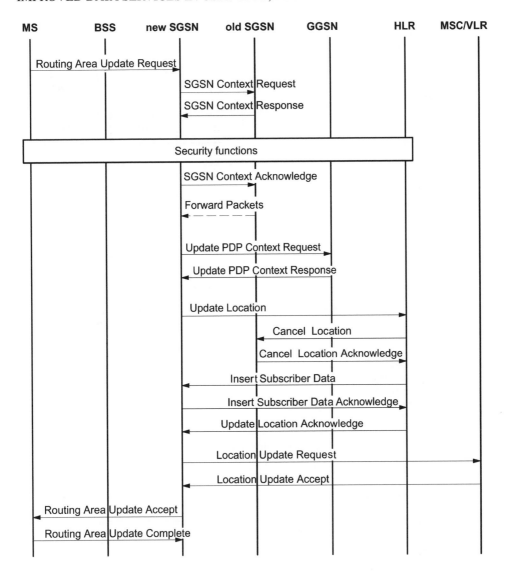

Figure 8.6 Inter-SGSN routing area update.

detection and error correction). An application running in the GPRS-MS (e.g. a browser) uses IP or X.25 in the network layer.

GPRS backbone: SGSN–GGSN. As mentioned earlier in this chapter, IP and X.25 packets are transmitted encapsulated within the GPRS backbone network. This is done using the GTP, i.e. GTP packets carry the user's IP or X.25 packets. GTP is defined both between

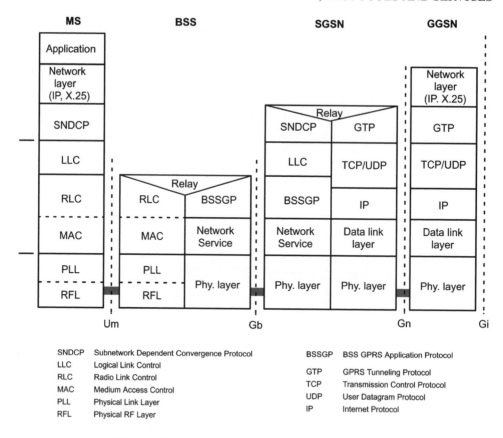

Figure 8.7 Protocol architecture: transmission plane.

GSNs within the same PLMN (Gn interface) and between GSNs of different PLMNs (Gp interface).

GTP contains procedures in the transmission plane as well as in the signaling plane. In the transmission plane, GTP employs a tunnel mechanism to transfer user data packets. In the signaling plane, GTP specifies a tunnel control and management protocol. The signaling is used to create, modify and delete tunnels. A Tunnel Identifier (TID), which is composed of the IMSI of the user and a Network Layer Service Access Point Identifier (NSAPI), uniquely indicates a PDP context. Below GTP, the standard protocols Transmission Control Protocol (TCP) or User Datagram Protocol (UDP) are employed to transport the GTP packets within the backbone network. TCP is used for X.25 (since X.25 expects a reliable end-to-end connection) and UDP is used for access to IP-based networks (which do not expect reliability in the network layer or below). In the network layer, IP is employed to route the packets through the backbone. Ethernet, ISDN or Asynchronous Transfer Mode (ATM)-based protocols may be used below IP. To summarize, in the GPRS backbone we have an IP/X.25-over-GTP-over-UDP/TCP-over-IP protocol architecture.

Air interface.   In the following we consider the air interface (Um) between MS and BSS or SGSN, respectively.

Subnetwork dependent convergence protocol.   The Subnetwork Dependent Convergence Protocol (SNDCP) is used to transfer packets of the network layer (IP and X.25 packets) between the MSs and their SGSN. Its functionality includes:

- multiplexing of several PDP contexts of the network layer onto one virtual logical connection of the underlying Logical Link Control (LLC) layer; and

- segmentation of network layer packets onto one frame of the underlying LLC layer and reassembly on the receiver side.

Moreover, SNDCP offers compression and decompression of user data and redundant header information (e.g. TCP/IP header compression).

Data link layer.   The data link layer is divided into two sublayers:

- LLC layer (between MS and SGSN); and

- RLC/Medium Access Control (MAC) layer (between MS and BSS).

The LLC layer provides a reliable logical link between a MS and its assigned SGSN. Its functionality is based on the LAPDm protocol (which is a protocol similar to HDLC and has been explained in section 5.3.1). LLC includes in-order delivery, flow control, error detection, retransmission of packets (ARQ) and ciphering functions. It supports variable frame lengths and different QoS classes, and besides point-to-point also point-to-multipoint transfer is possible. A logical link is uniquely addressed with a Temporary Logical Link Identifier (TLLI). Within one RA the mapping between TLLI and IMSI is unique. However, the user's identity remains confidential, since the TLLI is derived from the P-TMSI of the user.

The RLC/MAC layer has two functions. The purpose of the RLC layer is to establish a reliable link between the MS and the BSS. This includes the segmentation and reassembly of LLC frames into RLC data blocks and ARQ of uncorrectable blocks. The MAC layer controls the access attempts of MSs on the radio channel. It is based on a slotted-aloha principle (section 4.1). The MAC layer employs algorithms for contention resolution of access attempts, statistical multiplexing of channels and a scheduling and prioritizing scheme, which takes into account the negotiated QoS. On the one hand, the MAC protocol allows that a single MS simultaneously uses several physical channels (several time slots of the same TDMA frame). On the other hand, it also controls the statistical multiplexing, i.e. it controls how several MSs can access the same physical channel (the same time slot of successive TDMA frames). This is explained in more detail in section 8.1.7.

Physical layer.   The physical layer between MS and BSS can be divided into the two sublayers: Physical Link Layer (PLL) and Physical RF Layer (RFL). The PLL provides a physical channel between the MS and the BSS. Its tasks include channel coding (i.e. detection of transmission errors, forward error correction and indication of uncorrectable codewords),

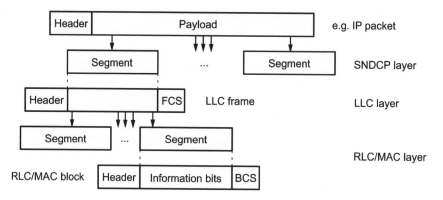

Figure 8.8 Data flow and segmentation between the protocol layers in the MS.

interleaving and detection of physical link congestion. The RFL, which operates below the PLL, includes modulation and demodulation.

To summarize this section, Figure 8.8 illustrates the data flow between the protocol layers in the MS. Packets of the network layer (e.g. IP packets) are passed down to the SNDCP layer, where they are segmented to LLC frames. After adding header information and a Frame Check Sequence (FCS) for error protection, these frames are segmented into one or several RLC data blocks. Those are then passed down to the MAC layer. One RLC/MAC block contains a MAC and RLC header, the RLC payload ('information bits') and a Block Check Sequence (BCS) at the end. The channel coding of RLC/MAC blocks and the mapping to a burst in the physical layer are explained in section 8.1.7.

BSS–SGSN interface.   At the Gb interface, the BSS GPRS Application Protocol (BSSGP) is defined on Layer 3. It is derived from the BSSMAP, which has been explained in section 5.3.1. The BSSGP delivers routing and QoS-related information between BSS and SGSN. The underlying Network Service (NS) protocol is based on the frame relay protocol.

Routing and conversion of addresses

Now we explain the routing example of section 8.1.3 in detail. Figure 8.9 roughly illustrates the transfer of an incoming IP packet. It arrives at the GGSN, is then routed through the GPRS backbone to the responsible SGSN and finally to the MS. Using the PDP context, the GGSN determines from the IP destination address a TID and the IP address of the relevant SGSN. Between GGSN and the SGSN, the GTP is employed. The SGSN derives the TLLI from the TID and finally transfers the IP packet to the MS. The so-called NSAPI is part of the TID. It maps a given IP address to the corresponding PDP context. An NSAPI/TLLI pair is unique within one RA. Figure 8.10 gives a similar example with an outgoing (mobile originated) IP packet.

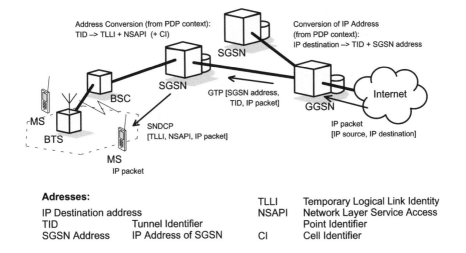

Figure 8.9 Routing and address conversion: incoming IP packet (mobile-terminated data transfer).

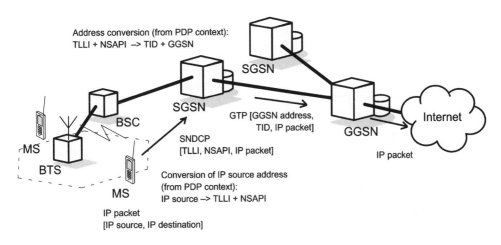

Figure 8.10 Routing and address conversion: outgoing IP packet (mobile-originated data transfer).

8.1.5 Signaling plane

The protocol architecture of the signaling plane comprises protocols for control and support of the functions of the transmission plane, e.g., for the execution of GPRS attach and detach, PDP context activation, the control of routing paths and the allocation of network resources.

Between MS and SGSN (Figure 8.11), the GPRS Mobility Management and Session Management (GMM/SM) protocol is responsible for mobility and session management.

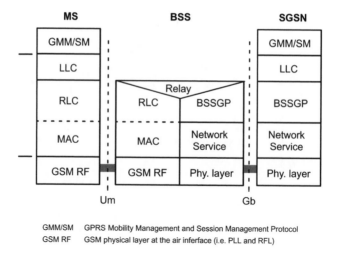

Figure 8.11 Signaling plane: MS-SGSN.

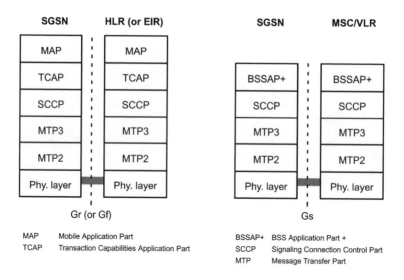

Figure 8.12 Signaling plane: SGSN-HLR, SGSN-EIR and SGSN-MSC/VLR.

It includes functions for GPRS attach/detach, PDP context activation, routing area updates and security procedures.

The signaling architecture between SGSN and the registers HLR, VLR and EIR (Figure 8.12) uses protocols known from conventional GSM (section 5.3) and partly extends them with GPRS-specific functionality. Between SGSN and HLR as well as between SGSN and EIR, an enhanced MAP is employed. The exchange of MAP messages is accomplished over the TCAP, the SCCP and the MTP.

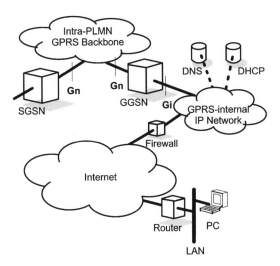

Figure 8.13 GPRS–Internet interconnection.

The BSS Application Part (BSSAP+) includes functions of GSM's BSSAP. It is applied to transfer signaling information between the SGSN and the VLR (Gs interface). This includes, in particular, signaling of the mobility management when coordination of GPRS and conventional GSM functions is necessary (e.g. for combined GPRS and nonGPRS location update, combined GPRS/IMSI attach or paging of a MS via GPRS for an incoming GSM call).

8.1.6   Interworking with IP networks

Figure 8.13 gives an example of how a GPRS network is interconnected with the Internet. From outside, i.e. from an external IP network's point of view, the GPRS network looks like any other IP subnetwork and the GGSN looks like a usual IP router.

As explained in section 8.1.3, each MS obtains an IP address after its GPRS attach, which is valid for the duration of the session. The network provider has reserved a certain number of IP addresses, and can dynamically assign these addresses to active MSs. To do so, the network provider may install a Dynamic Host Configuration Protocol (DHCP) server in its network. This server automatically manages the available address space. The address resolution between IP address and GSM address is performed by the GGSN, using the appropriate PDP context. The routing of IP packets and the tunneling through the intra-PLMN backbone (using GTP) has been explained in sections 8.1.1, 8.1.3 and 8.1.4.

Moreover, a Domain Name Server (DNS) is used to map between IP addresses and host names. To protect the PLMN from unauthorized access, a firewall is installed between the private GPRS network and the external IP network.

With this configuration, GPRS can be seen as a wireless extension of the Internet all the way to a MS. The mobile user has direct connection to the Internet.

8.1.7   Air interface

The enhanced air interface of GPRS offers higher data rates and a packet-oriented transmission. It is therefore considered one of the key aspects in GPRS. In this section, we explain how several MSs can share one physical channel (multiple access) and how the assignment of radio resources between circuit-switched GSM services and GPRS services is controlled. Afterwards, the logical channels and their mapping onto physical channels (using multiframes) is presented. Finally, GPRS channel coding concludes this chapter.

Multiple access and radio resource management

On the physical layer, GPRS uses the GSM combination of FDMA and TDMA with eight time slots per TDMA frame (as explained in section 4.2.2). However, several new methods are used for channel allocation and multiple access. They have a significant impact on the performance of GPRS.

In circuit-switched GSM, a physical channel (i.e. one time slot of successive TDMA frames) is permanently allocated for a particular MS during the entire call period (no matter whether data are transmitted or not). Moreover, it is assigned in the uplink as well as in the downlink.

GPRS enables a far more flexible resource allocation scheme for packet transmission. A GPRS MS can transmit on several of the eight time slots within the same TDMA frame (multislot operation). The number of time slots which a MS is able to use is called a multislot class. In addition, uplink and downlink are allocated separately, which saves radio resources, especially for asymmetric traffic (e.g. Web browsing).

A cell supporting GPRS must allocate physical channels for GPRS traffic. In other words, the radio resources of a cell are shared by all MSs (GSM and GPRS) located in this cell. The mapping of physical channels to either GPRS or circuit-switched GSM services can be performed in a dynamic way. A physical channel which has been allocated for GPRS transmission is denoted as a Packet Data Channel (PDCH). The number of PDCHs can be adjusted according to the current traffic demand (capacity on demand principle). For example, physical channels not currently in use by GSM calls can be allocated as PDCHs for GPRS to increase the QoS for GPRS. When there is a resource demand for GSM calls, PDCHs may be deallocated.

As already mentioned, physical channels for packet-switched transmission (PDCHs) are only allocated for a particular MS when this MS sends or receives data packets, and they are released after the transmission. With this dynamic channel allocation principle, multiple MSs can share one physical channel. For bursty traffic this results in a much more efficient usage of the radio resources.

The channel allocation is controlled by the BSC. To prevent collisions, the network indicates which channels are currently available in the downlink. An Uplink State Flag (USF) in the header of downlink packets shows which MS is allowed to use this channel in the uplink. The allocation of PDCHs to a MS also depends on its multislot class and the QoS of the session.

Table 8.3 Logical channels in GPRS.

Group		Channel	Function	Direction
Traffic channels	Packet data traffic channel	PDTCH	Packet data traffic	MS ↔ BSS
Signaling channels	Packet broadcast control channel	PBCCH	Packet broadcast control	MS ← BSS
	Packet common control channel (PCCCH)	PRACH	Packet random access	MS → BSS
		PAGCH	Packet access grant	MS ← BSS
		PPCH	Packet paging	MS ← BSS
		PNCH	Packet notification	MS ← BSS
	Packet dedicated control channels	PACCH	Packet associated control	MS ↔ BSS
		PTCCH	Packet timing advance control	MS ↔ BSS

Logical channels

Table 8.3 lists the packet data logical channels defined in GPRS. As with logical channels in conventional GSM, they can be divided into two categories: traffic channels and signaling (control) channels. The signaling channels can further be divided into packet broadcast control, packet common control, and packet dedicated control channels.

The Packet Data Traffic Channel (PDTCH) is employed for the transfer of user data. It is assigned to one MS (or, in the case of PTM, to multiple MSs). One MS can use several PDTCHs simultaneously.

The Packet Broadcast Control Channel (PBCCH) is a unidirectional point-to-multipoint signaling channel from the BSS to the MSs. It is used by the BSS to broadcast information about the organization of the GPRS radio network to all GPRS MSs of a cell. In addition to system information about GPRS, the PBCCH should also broadcast important system information about circuit-switched services, so that a GSM/GPRS MS does not need to listen to the BCCH.

The Packet Common Control Channel (PCCCH) transports signaling information for functions of the network access management, i.e. for allocation of radio channels, medium access control and paging. Four sub-channels are defined:

- the Packet Random Access Channel (PRACH) is used by the MSs to request one or more PDTCH;

- the Packet Access Grant Channel (PAGCH) is used to allocate one or more PDTCH to a MS;

- the Packet Paging Channel (PPCH) is used by the BSS to find the location of a MS (paging) prior to downlink packet transmission;

- the Packet Notification Channel (PNCH) is used to inform MSs of incoming PTM messages.

Figure 8.14 shows the principle of the uplink channel allocation (mobile-originated packet transfer). A MS requests a channel by sending a `PACKET CHANNEL REQUEST` on the

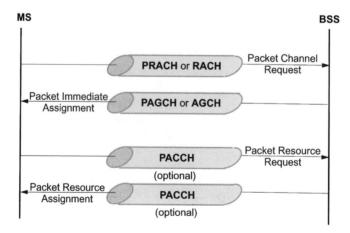

Figure 8.14 Uplink channel allocation (mobile-originated packet transfer).

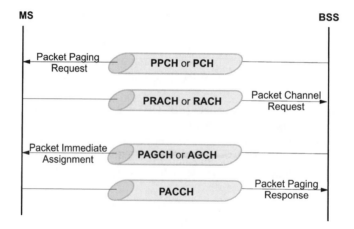

Figure 8.15 Paging (mobile-terminated packet transfer).

PRACH or RACH. The BSS answers on the PAGCH or AGCH, respectively. Once the PACKET CHANNEL REQUEST is successful, a so-called Temporary Block Flow (TBF) is established. With that, resources (e.g. PDTCH and buffers) are allocated for the MS, and data transmission can start. During transfer, the USF in the header of downlink blocks indicates to other MSs that this uplink PDTCH is already in use. On the receiver side, a Temporary Flow Identifier (TFI) helps to reassemble the packet. Once all data has been transmitted, the TBF and the resources are released again. Figure 8.15 illustrates the paging procedure of a mobile station (mobile-terminated packet transfer).

The packet dedicated control channel is a bidirectional point-to-point signaling channel. It contains the following channels.

Table 8.4 Logical channels in GPRS.

Channel type	Net data throughput (kbit/s)	Block length (bits)	Block distance (ms)
PDTCH (CS-1)	9.05	181	–
PDTCH (CS-2)	13.4	268	–
PDTCH (CS-3)	15.6	312	–
PDTCH (CS-4)	21.4	428	–
PACCH	Changes dynamically		
PBCCH	$s \times 181/120$	181	120
PAGCH	Changes dynamically	181	–
PNCH	Changes dynamically	181	–
PPCH	Changes dynamically	181	–
PRACH (8 bit Access burst)	Changes dynamically	8	–
PRACH (11 bit Access burst)	Changes dynamically	11	–

- The Packet Associated Control Channel (PACCH) is always allocated in combination with one or more PDTCH. It transports signaling information related to one specific MS (e.g. power control information).

- The Packet Timing Advance Control Channel (PTCCH) is used for adaptive frame synchronization. The MS sends over the uplink part of the PTCCH, the PTCCH/U, ABs to the BTS. From the delay of these bursts, the correct value for the TA can be derived; see section 4.3.2. This value is then transmitted in the downlink part, the PTCCH/D, to inform the MS.

Coordination between circuit-switched and packet-switched logical channels is also possible here. If the PCCCH is not available in a cell, a GPRS MS can use the CCCH of circuit-switched GSM to initiate the packet transfer. Moreover, if the PBCCH is not available, it can obtain the necessary system information via the BCCH.

Table 8.4 lists the block lengths and net data throughput of the logical GPRS channels (compare with Table 4.2). Four different coding schemes (CS-1 to CS-4) are defined for data transmission on the PDTCH. They are explained in section 8.1.7.

As with circuit-switched GSM, the GPRS logical channels can be used in certain combinations only. The allowed combinations for multiplexing logical channels onto physical channels are shown in Table 8.5. Moreover, Table 8.6 shows the channel configurations which a GPRS MS can use (dependent on its state). Combination M9 represents a MS in an IDLE state waiting for incoming packets. Combination M10 is a transmitting MS with multislot capabilities. Several PDTCHs are assigned to a single MS, where n denotes the number of PDTCHs which allow bidirectional transmission, and m denotes the number of PDTCHs which allow only unidirectional transmission. We have: $n = 1, \ldots, 8$, $m = 0, \ldots, 8$ and $n + m = 1, \ldots, 8$.

Table 8.5 Combinations of logical GPRS channels.

	B10	B11	B12	B13
PDTCH	■	■	■	
PBCCH	■			■
PCCCH	■	■		■
PACCH	■	■	■	
PTCCH	■	■	■	

Table 8.6 Channel combinations used by the MS.

	M9	M10
PDTCH		$n+m$
PBCCH	■	
PCCCH	■	
PACCH		■
PTCCH		■

Mapping of packet data logical channels onto physical channels

From section 4.4 we know that the mapping of logical GSM channels onto physical channels has two components: mapping in frequency and mapping in time. The mapping in frequency is based on the TDMA frame number and the frequencies allocated to the BTS and the MS. The mapping in time is based on the definition of complex multiframe structures on top of the TDMA frames.

A multiframe structure for PDCHs consisting of 52 TDMA frames (each with eight time slots) is shown in Figure 8.16. The corresponding time slots of a PDCH of four consecutive TDMA frames form one radio block (blocks B0–B11). Two TDMA frames are reserved for transmission of the PTCCH, and the remaining two frames are IDLE frames. A multiframe has thus a duration of approximately 240 ms (52 × 4.615 ms). A radio block consists of 456 bits.

The mapping of the logical channels onto the blocks B0–B11 of the multiframe can vary from block to block and is controlled by parameters which are broadcast on the PBCCH. The GPRS recommendations define which time slots may be used by a logical channel.

In addition to the 52-multiframe structure, which can be used by all logical GPRS channels, a 51-multiframe structure is also defined. It is used for PDCHs carrying only the logical channels PCCCH and PBCCH (channel combination B13 in Table 8.5). In the downlink, it consists of 10 blocks each of 4 frames (B0-B9) and 10 IDLE frames. In the uplink, it has 51 random access frames. Its duration is 235.4 ms.

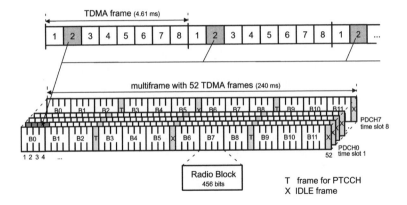

Figure 8.16 Multiframe structure with 52 TDMA frames.

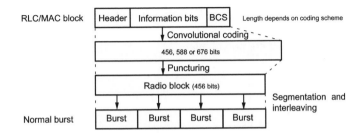

Figure 8.17 Physical layer at the air interface: channel coding, interleaving and formation of bursts (continued from Figure 8.8).

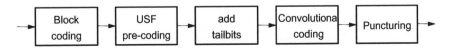

Figure 8.18 Encoding of GPRS data blocks.

Channel coding

Figure 8.17 shows how a block of the RLC/MAC layer (compare with Figure 8.8) is encoded and mapped onto four bursts. Channel coding is used to protect the transmitted data packets against errors and perform forward error correction. The channel coding technique in GPRS is quite similar to that employed in conventional GSM. An outer block coding, an inner convolutional coding and an interleaving scheme are used (Figure 4.29).

Four coding schemes with different code rates are defined. Their parameters are listed in Table 8.7. For each scheme, a block of 456 bits results after encoding. Figure 8.18 illustrates the encoding process, which is briefly explained in the following.

Table 8.7 Channel coding schemes for the traffic channels in GPRS.

Coding scheme	Pre-encoded USF	Infobits without USF and BCS	Parity bits BCS	Tail bits	Output convolutional encoder	Punctured bits	Code rate	Data rate (kbit/s)
CS-1	3	181	40	4	456	0	1/2	9.05
CS-2	6	268	16	4	588	132	≈ 2/3	13.4
CS-3	6	312	16	4	676	220	≈ 3/4	15.6
CS-4	12	428	16	–	456	–	1	21.4

Let us employ coding scheme CS-2. First of all, the 271 information bits of an RLC/MAC block (268 bits plus 3 bits USF; Table 8.4) are mapped to 287 bits using a systematic block encoder, i.e. 16 parity bits are added. These parity bits are denoted as a BCS. The USF pre-encoding maps the first 3 bits of the block (i.e. the USF) to 6 bits in a systematic way. Afterwards, 4 zero bits (tail bits) are added at the end of the entire block. The tail bits are needed for the termination of the subsequent convolutional coding.

For the convolutional coding, a nonsystematic rate-1/2 encoder with memory 4 is used, which is defined by the generator polynomials

$$G_0(d) = 1 + d^3 + d^4,$$

$$G_1(d) = 1 + d + d^3 + d^4.$$

This is the same encoder as used in conventional GSM. A possible encoder realization is shown in Figure 4.32. At the output of the convolutional encoder a codeword of length 588 bits results. Following this, 132 bits are punctured, resulting in a radio block of length 456 bits. Thus, we obtain a code rate of the convolutional encoder (including the puncturing) of

$$r = \frac{6 + 268 + 16 + 4}{456} \approx \frac{2}{3}.$$

Coding scheme CS-1 is equivalent to the coding of the SACCH. A systematic fire code is used for block coding (see section 4.8.1, first paragraph). There is no pre-coding of the USF bits. The convolutional coding is performed with the known rate-1/2 encoder, however, this time the output sequence is not punctured. Using CS-4, the 3 USF bits are mapped to 12 bits, and no convolutional coding is applied.

For the coding of the traffic channels (PDTCH), one of the four coding schemes is chosen, depending on the quality of the signal. The two SFs in a NB (Figure 4.7) are used to indicate which coding scheme is used. Under very bad channel conditions, CS-1 yields a data rate of only 9.05 kbit/s per time slot, but a very reliable coding. Under good channel conditions, convolutional coding is skipped (CS-4), and we achieve a data rate of 21.4 kbit/s per time slot. Thus, we obtain a theoretical maximum data rate of 171.2 kbit/s per TDMA frame. In practice, multiple users share the time slots and, thus, a much lower bit rate is available to the individual user. Moreover, the quality of the radio channel will not always allow us to use CS-4 (or CS-4 is not supported by the mobile terminal or by the network operator). The data rate available to the user depends (among other things) on the current total traffic load in the cell (i.e. the number of users and their traffic characteristics), the used coding scheme, and the multislot class of the MS. Data rates between 10 and 50 kbit/s are realistic values. A simulative study on GPRS performance can be found in Kalden *et al.* (2000).

After encoding, the codewords are input into a block interleaver of depth 4. For all coding schemes, the interleaving scheme known from the interleaving of the SACCH (see section 4.8.3, last paragraph) is employed. On the receiver side, the codewords are deinterleaved. As in GSM, the decoding is performed using the Viterbi algorithm.

The signaling channels are encoded using CS-1. An exception is the PRACH. It can transmit two very short bursts, one burst with 8 information bits and one burst with 11 information bits. The coding for the 8-bit burst is the one used for the RACH (see sections 4.8.1 and 4.8.2) and the coding for the 11-bit burst is a punctured version of it.

8.1.8 Authentication and ciphering

The security principles inside the GPRS network are almost equivalent to those used in conventional GSM (section 5.6). Security functions in the GPRS network:

- protect against unauthorized use of services (by authentication and service request validation);

- provide data confidentiality (using ciphering); and

- provide confidentiality of the subscriber identity.

As in GSM, two keys are used: the subscriber authentication key Ki and the cipher key Kc. The main difference is that the SGSN, not the MSC, which handles authentication. Moreover, a special GPRS ciphering algorithm (A5) has been defined, which is optimized for encryption of packet data.

User authentication

Figures 8.19 and 8.20 illustrate the GPRS authentication process. The standard GSM algorithms are used to generate security data. The algorithm A3 calculates the signature response (SRES) from the subscriber authentication key (Ki) and a random number (RAND).

If the SGSN does not have authentication sets for a user (Kc, RAND, SRES), it requests them from the HLR by sending a message SEND AUTHENTICATION INFO. The HLR responds with a SEND AUTHENTICATION INFO ACK which includes the security data. Now, the SGSN offers a random number RAND to the MS (AUTHENTICATION AND CIPHERING REQUEST). The MS calculates SRES and transmits it back to the SGSN (AUTHENTICATION AND CIPHERING RESPONSE). If the SRES of the MS is equal to the SRES calculated (or maintained) by the SGSN, the user is authenticated and is allowed to use the network.

Ciphering

The ciphering functionality is performed in the LLC layer between MS and SGSN (see Figures 8.7 and 8.11). Thus, the ciphering scope reaches from the MS all the way to the SGSN (and vice versa), whereas in conventional GSM the scope is only between MS and BTS/BSC.

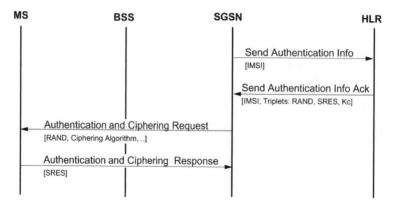

Figure 8.19 Subscriber authentication in GPRS.

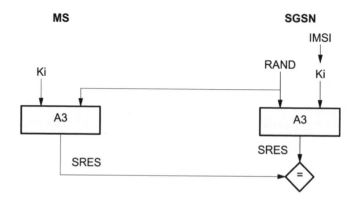

Figure 8.20 Principle of subscriber authentication in GPRS.

As in GSM ciphering, the algorithm A8 generates the cipher key Kc from the key Ki and a random number RAND (see Figure 5.53). Kc is then used by the GPRS Encryption Algorithm (GEA) for data encryption (algorithm A5). Note that the key Kc which is handled by the SGSN is independent of the key Kc handled by the MSC for conventional GSM services. A MS may thus have more than one Kc key.

The MS and the SGSN start ciphering after the message AUTHENTICATION AND CIPHERING RESPONSE is sent or received, respectively. Afterwards, GPRS user data and signaling during data transfer are transmitted in an encrypted manner.

Subscriber identity confidentiality

As in GSM, the identity of the subscriber is confidential. This is done by using temporary identities on the radio channel. In particular, the user's IMSI is not transmitted unencrypted, instead a Packet Temporary Mobile Subscriber Identity (P-TMSI) is assigned to each user by the SGSN. This address is temporary and is only valid and unique in the service area of this

SGSN. From the P-TMSI, a TLLI can be derived. The mapping between these temporary identities and the IMSI is stored only in the MS and in the SGSN.

8.1.9 Summary of GPRS

GPRS has been an important step in the evolution of cellular networks towards 3G and mobile Internet. Its packet-oriented transmission technology enables efficient and simplified wireless access to IP networks.

GPRS extends the existing GSM infrastructure in particular with two network nodes, namely the SGSN and GGSN. In section 8.1.1 their tasks and the interworking with GSM nodes and registers (HLR, VLR and EIR) has been explained.

An important feature of GPRS is its QoS support. An individual QoS profile (service precedence, reliability, delay, and throughput) can be negotiated for each PDP context. For the simultaneous use of GPRS and conventional GSM services, three classes of MSs are defined in the standard.

Before a GPRS MS can use GPRS services it must obtain an address used in the external packet data network (e.g. an IP address) and create a PDP context. This context describes the essential characteristics of the session (PDP type, PDP address, QoS, and GGSN). In order to support a large number of mobile users, it is essential to use dynamic address allocation, e.g., using DHCP for dynamic IP address assignment.

Once a MS has an active PDP context, packets addressed from the external packet data network to the MS will be routed to the responsible GGSN. The GGSN then tunnels them to the current SGSN of the mobile user, which finally forwards the data to the MS. The GPRS location management is based on the definition of a MS state model. Depending on the state of the MS (READY, STANDBY, or IDLE), it performs many or only few location updates. For this purpose, special routing areas are defined, which are sub-areas of the location areas defined in GSM. Although GPRS has its own mobility management, it cooperates with the GSM mobility management. This results, for example, in a more efficient paging mechanism for MSs that use circuit- and packet-based services simultaneously.

In section 8.1.4 we showed the protocol architecture of the GPRS transmission and signaling plane. GPRS-specific protocols include GTP, the GMM/SM protocol, and SNDCP. Some GSM protocols, such as MAP, have been extended for use with GPRS.

The packet-oriented air interface is one of the key aspects of GPRS. MSs with multislot capability can transmit on several time slots of a TDMA frame, uplink and downlink are allocated separately, and physical channels are only assigned for the duration of the transmission, which leads to a statistical multiplex gain. This flexibility in the channel allocation results in a more efficient utilization of the radio resources. On top of the physical channels, a number of logical packet channels have been standardized. The traffic channel PDTCH is used for payload transmission. The GPRS signaling channels are used, e.g., for broadcast of system information (PBCCH), access control (PRACH, PAGCH), paging (PPCH), and notification of incoming PTM messages (PNCH). Again, the coordination between GPRS and GSM channels saves radio resources.

GPRS channel coding defines four different coding schemes, which allows the tradeoff between the level of error protection and data rate to be adjusted, depending on the current radio channel quality. GPRS security principles include authentication, ciphering, and subscriber identity confidentiality. The SGSN handles authentication, and a special GEA has been defined. Moreover, GPRS operators protect their network with firewalls to external networks and border gateways to other GPRS networks. IP security protocols (IPsec) may be used to communicate over insecure external IP networks.

Typical scenarios for GPRS are wireless access to the Internet, e-mail communication, WAP over GPRS, and applications in the telemetry field. Users can access the Internet without first requiring to dial into an Internet service provider. In particular, mobile e-commerce and location-based services (e.g. tourist guides) have gained importance in the last few years. The main advantages for the users are the higher data rates and volume-based billing. The latter allows them to stay online for a long time.

8.2 HSCSD

As the name implies, the High Speed Circuit Switched Data Service is, in contrast to GPRS, circuit-switched. That is, the user has a fixed data rate bearer available for the duration of the data connection. This is independent of the amount of data actually transmitted, such that the connection has to be paid for, even during periods, in which no data are transmitted, according to the respective higher layer services used. This means that HSCSD is specifically useful for applications which demand for a fixed data rate. The advantage of this, however, is that the data rate is guaranteed during the connection time, in case a transparent bearer service is applied. On the other hand, the QoS is secured, if a nontransparent bearer service is applied. HSCSD supports both options.

Just as in GPRS, also HSCSD allows for the parallel use of several, say n, traffic channels to provide higher data rates. Figure 8.21 depicts an example with $n = 2$, in which timeslots TS1 and TS2 are used for one HSCSD connection, both in uplink and downlink. The principal restriction for the number of timeslots n is given by the requirement that they all have to reside on the same frequency channel. Therefore, the standard allows for up to $n = 8$ timeslots or channels to be assigned to one user. This gives us the maximal data rate achievable: using eight channels at once, each carrying a full-rate traffic channel TCH/F9.6, a sum data rate of 76.8 kbit/s could be achieved. However, it should be recalled that, even though GSM applies FDD, uplink and downlink timeslots of the same TCH have an offset, such that the mobile terminal can perform transmission and reception subsequently. This because terminals with the ability to perform transmit and receive operations in parallel would imply much higher complexity at the terminal and thus would make terminals much more expensive. Therefore, earlier versions of the HSCSD standard only allowed for up to $n = 4$ timeslots to be used at once. This would present us with a data rate of 38.4 kbit/s, when four TCH/F9.6 channels are used. In fact, by applying a different coding scheme, a maximum data rate of 57.6 kbit/s is achieved with $n = 4$. Figure 8.21 shows, for the example of $n = 2$, how the required operations in the mobile terminal can still be performed sequentially, as was originally intended in classical GSM: the mobile is receiving in timeslots TS1 and TS2 of the downlink frame. Owing to the time offset of three timeslots between the uplink and the downlink frames, the terminal has already completed reception, when it has to start transmitting on the uplink, also using TS1 and TS2, but now on the uplink frames. After completion of transmission, and before having to receive data again in the following TS1 and TS2 on the downlink, there is still enough time to monitor BCCH carriers of neighboring cells, which is important, e.g., for handover issues.

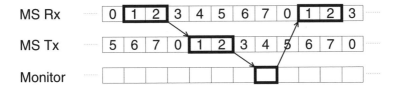

Figure 8.21 Example HSCSD channel occupation with $n = 2$.

When assigning timeslots, the HSCSD service takes the so-called multislot class of the MS into consideration. A list of the multislot classes considered for HSCSD is shown in Table 8.8. It lists the class name, the maximum allowed number of downlink timeslots (Rx), uplink timeslots (Tx), and sum of downlink and uplink timeslots. For instance, multislot class 3 terminals might use up to two timeslots for reception (downlink) and up to two timeslots for transmission (uplink). However, the sum of uplink and downlink timeslots must never be larger than three. When a mobile of a certain multislot class initiates a HSCSD connection, the RRM must consider the restrictions of the respective multislot class, when assigning timeslots. The last column in Table 8.8 refers to the type of the MS: type 1 terminals do not have the ability to transmit and receive at the same time. Therefore, the RRM must ensure the appropriate selection of timeslots. In contrast, type 2 mobiles do have the ability to transmit and receive in parallel, which gives the RRM more options to allocate timeslots for HSCSD.

8.2.1 Architecture

From an architectural point of view, HSCSD does not demand many changes as compared with standard GSM data services (see Appendix A). The GSM architecture for HSCSD support is depicted in Figure 8.22. At the U_m interface between MS and BTS up to $n = 8$ TCH/F channels are used, which are forwarded transparently via the A_{bis} interface from the BTS to the BSC. Here the data from all parallel channels are multiplexed on a single 64 kbit/s connection and transmitted over the A interface to the MSC. The n full rate channels are considered independent of each other and are treated individually for the purpose of, e.g., air interface error control. However, logically they belong to the same HSCSD configuration and are controlled as one radio link by the network for the purpose of cellular operations, such as handover. This requires new BSS functionality.

The main difference to the standard GSM data services is in a combining and splitting functionality that is demanded at the MS and the MSC, combining and splitting, respectively, the multiple data streams that are transmitted between both entities. This functionality is provided in the Terminal Adoption Function (TAF) at the MS and in the IWF at the MSC, respectively (Figure 8.22).

8.2.2 Air interface

As described above at the air interface, a HSCSD connection comprises n traffic channels (TCH). All n channels use the same hopping sequences, if frequency hopping is applied. Also the training sequence assigned, is the same for all n traffic channels. However, each traffic channel is assigned an independent SACCH signaling channel, allowing for

Table 8.8 HSCSD MS multislot classes.

Multislot class	Maximum number of slots			
	Rx	Tx	Sum	Type
1	1	1	2	1
2	2	1	3	1
3	2	2	3	1
4	3	1	4	1
5	2	2	4	1
6	3	2	4	1
7	3	3	4	1
8	4	1	5	1
9	3	2	5	1
10	4	2	5	1
11	4	3	5	1
12	4	4	5	1
13	3	3	NA	2
14	4	4	NA	2
15	5	5	NA	2
16	6	6	NA	2
17	7	7	NA	2
18	8	8	NA	2

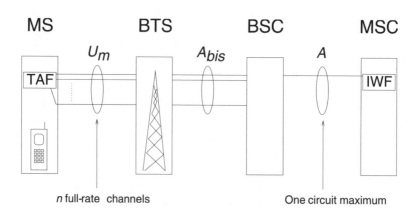

Figure 8.22 GSM architecture for HSCSD support.

independent signal level and quality reporting on each timeslot, in turn, allowing for individual power control on each timeslot. This is done, since the interference level can be different on each timeslot such that different power levels might be required in order to secure the required signal quality on every single timeslot. The HSCSD connection has just one FACCH control channel assigned for management purposes. For all HSCSD channels, the same channel coding is applied, however, for nontransparent services, different channel codings can be used for uplink and downlink, respectively. For each HSCSD channel, different ciphering keys are used, derived from Kc. The channel assignment configuration can be either symmetric or asymmetric and channels can be allocated either on consecutive or on nonconsecutive timeslots of the same carrier, as long as the restrictions of the given multislot class of the MS are considered.

Symmetric connections comprise a bidirectional FACCH as well as co-allocated bi-directional TCH/F and SACCH channels. In contrast, asymmetric connections can have unidirectional and/or bidirectional TCH/f and SACCH channels in addition to the bi-directional FACCH. Unidirectional channels in HSCSD are always downlink channels only. Both symmetric and asymmetric HSCSD channels have a bidirectional main channel, which carries the FACCH.

8.2.3 HSCSD resource allocation and capacity issues

When initiating a HSCSD connection, the user indicates during setup the maximum number of traffic channels, the multislot class of the terminal, acceptable channel codings, and a wanted fixed network user rate. In the case of a nontransparent connection, the wanted air interface user rate is also indicated. The connection requirements can be symmetric or asymmetric, considering uplink and downlink. These parameters are then used by the network to allocate appropriate resources and to setup the demanded HSCSD connection. The minimum channel requirement is always one TCH/F. This means that transparent and nontransparent connections can be established with any number of TCH/F from one up to the specified maximum number. When the user rate requirements cannot be met, the network will give priority to fulfill the air interface user requirement in downlink direction, possibly using an asymmetric configuration. The network can use dynamic resource allocation for nontransparent HSCSD connections, as long as the allocated channel configurations are always in line with the limiting values defined by the MS and with the multislot class of the terminal. In the case of transparent HSCSD connections, dynamic resource allocation is only allowed if the air interface user data rate is kept constant. The network performs the change of channel allocation configurations during the HSCSD connection, by means of resource upgrading and resource downgrading procedures. If indicated by the MS during call setup, the MS may issue a request an upgrade or downgrade of service level anytime during the HSCSD connection.

Obviously, HSCSD can increase the data rate of a single user. This, however, is achieved at the cost of assigning more frequency resources to a single user. This implies that from the point of view of a network provider, resources are more frequently occupied. A single HSCSD user using $n = 4$ TCHs will occupy the resources of four potential voice connections. Therefore, when many users utilize HSCSD at once in a single cell, the blocking performance in that cell will deteriorate. This will influence the radio resource management strategy that a network provider will use: if possible, the provider might try to assign more frequency carriers to cells in areas where the frequent use of HSCSD seems likely, for instance at airports or in business areas of a city. Also, the provider has the option of restricting the number of HSCSD connections in a cell in favor of voice connections.

Another interesting aspect of the HSCSD radio resource management is the handover. When a handover of the HSCSD connection to a neighboring cell becomes necessary, it might not be possible to find n free TCHs on a single frequency channel in the target cell. In this case, two options are possible: either the HSCSD connection is simply dropped or the connection is resumed with a lower number of TCHs. For this purpose, the resource downgrading procedure is applied, as described above for the case of dynamic resource allocation. With this concept, the probability of a handover being blocked and thus a HSCSD call being dropped can be reduced. Later, a resource upgrading procedure can be applied if an appropriate resource becomes available, and the original channel allocation configuration can be recreated.

8.3   EDGE

As discussed above in this chapter, HSCSD and GPRS achieve higher data rates because a MS can use several time slots of the same TDMA frame and partly because new coding schemes are employed. The EDGE[2] system goes one step further, by improving the spectral efficiency on the physical layer on a single timeslot (Furuskär *et al.*, 1999). Technically speaking, EDGE can be considered mainly as an air interface improvement. However, in effect it is a system concept that is used in order to introduce new bearer services into GSM systems. In this context it is interesting to note that both GPRS and EDGE have also been standardized for the North American cellular network TDMA-136 (GPRS-136 and GPRS-136HS EDGE). Within the GSM context, EDGE is used to improve the existing data services with a focus on GPRS and HSCSD, which become Enhanced GPRS (EGPRS) and Enhanced Circuit Switched Data (ECSD), respectively, when combined with EDGE technology.

8.3.1   The EDGE concept

A classical GSM system is designed and planned in a worst-case fashion: the radio network planning is carried out such that there is a high probability that all users in the network will experience a minimum signal quality that is sufficient for low error probability with a fixed modulation and error coding scheme. In fact, the main restrictions for the radio network planning come from those users that are located at the cell edges, far away from base stations. So, it can be said that the GSM network is designed for the cell edge users. However, users located closer to the base stations are likely to experience signal quality levels that are much better than required for the standard GSM modulation and coding schemes. To this end, EDGE introduces several additional combinations of modulation and coding schemes, which allow terminals to adapt their data rates to their individual signal quality levels. For this purpose a link adaptation technique is introduced with EDGE, which dynamically chooses a modulation and coding scheme according to the current radio channel conditions.

---

[2]EDGE used to stand for Enhanced Data Rates for *GSM* Evolution, however, *GSM* was replaced with *Global*, since EDGE, as mainly an air interface enhancement, can be, and actually is, used not only in GSM but in various cellular networks with different standards.

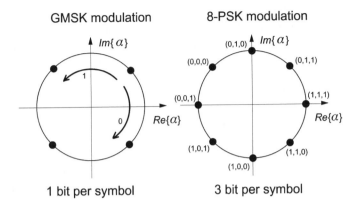

Figure 8.23 Symbol space constellations for GMSK and 8-PSK.

Figure 8.24 The EDGE burst has a training sequence of 26 symbols in the middle, three tail symbols at either end, and 8.25 guard symbols at the end. The burst carries 2 × 58 data symbols.

8.3.2 EDGE physical layer, modulation and coding

For EDGE, in addition to the GMSK modulation scheme used in GSM, an 8-Phase Shift Keying (8-PSK) scheme is available, which achieves an approximately three times higher data rate per time slot and hence a higher spectral efficiency. Using GMSK, one data bit d_i is mapped to one symbol a_i (see section 4.2.1); with 8-PSK, three data bits d_i are combined into one symbol a_i and transmitted together. Figure 8.23 shows the symbol constellations in the complex plane and the associated bit sequences. As opposed to GMSK, 8-PSK does not have a constant envelope and therefore puts higher requirements on new transceivers in BTSs and MSs.

In order to achieve compatibility with the GSM system, most EDGE physical layer parameters are the same as in GSM: the carrier spacing is 200 kHz and the TDMA frame structure remains unchanged. The burst format for the 8-PSK modulated transmission is also similar to the standard GSM frames (Figure 8.24): A burst comprises a 26-symbol training sequence as midamble, three tail symbols at the beginning and end of the frame, and 8.25 guard symbols at the end. Before and after the midamble, the frame carries 58 data symbols, each symbol representing three bits according to the 8-PSK modulation applied.

Several different coding schemes with different code rates can be combined with the two different modulation schemes available (Tables 8.9 and 8.10) (Molkdar *et al.*, 2002). In order to select the optimal scheme, EDGE applies a link adaptation based on the current channel quality.

In addition to the additional modulation and coding schemes, EDGE also introduces a code combining technique called incremental redundancy, also known as Hybrid Automatic Repeat Request (HARQ). In the incremental redundancy mode, the first RLC data block can be transmitted with some or even no redundancy added. If the data block cannot be decoded correctly, in the next retransmission, more redundancy is sent, applying a different puncturing scheme on the same RLC block. The erroneous blocks are stored, such that they can be combined with each retransmission required, until the RLC can be decoded correctly.

Also, the RLC/MAC layer has been enhanced with appropriate RLC/burst mapping functionality. The improved RLC/MAC layer allows for resegmentation to enable for retransmission with different coding schemes, independent coding of RLC/MAC headers and an increased ACK/NACK window size. As for the segmentation, each LLC PDU is broken into 20 ms RLC data blocks, to match the TDMA burst structure of GSM. Depending on the selected coding scheme by the link adaptation, the number of bits that fit into a RLC data block varies. The required RLC/MAC headers are appended to the user data before transmission. The content of the RLC/MAC headers includes a Block Sequence Number (BSN), and a Coding and Puncturing Scheme (CPS) indicator, that is required for the code combining process in the HARQ procedure. Then, check sequences for the user data (BCS) and for the header (HCS) are added to form a RLC radio block. The RLC radio block is then passed to the physical layer. There the user data and the header are coded separately and finally mapped to two or four TDMA bursts, depending on the coding scheme. The modulation and coding scheme can be changed for each RLC block, i.e. typically every four TDMA bursts. However, the modifications will be based on channel quality monitoring and changes will, in fact, be more seldom, depending on the channel measurement and reporting frequency.

8.3.3   EDGE: effects on the GSM system architecture

It is mainly the improved data rate of EDGE that imposes new requirements also for the GSM/GPRS network architecture: the main bottleneck for EDGE in GSM is the A_{bis} interface between BTS and BSC. In standard GSM, this interface supports only 16 kbit/s per traffic channel. However, EDGE can support close to 64 kbit/s for one traffic channel. Therefore, EDGE requires the allocation of multiple A_{bis} slots to one traffic channel. In future 3G network architectures, this requirement might also be fulfilled by applying ATM or IP-based solutions. The A interface between BSC and MSC can handle 64 kbit/s already in standard GSM and therefore does not present a problem. Other than that, the GSM/GPRS network architecture is not affected by the introduction of EDGE. This is due to the fact that EDGE is foremost just an enhanced air interface technology. Two different types of EDGE mobile terminals are considered in the standard: terminals that are capable of 8-PSK modulation on the downlink only, and terminals which provide 8-PSK capability on both the uplink and downlink. The first class of terminals can benefit from higher data rates on the downlink, which is still considered to be the link of higher importance, due to the popularity of services which are heavy on the downlink data amount, such as browsing and file downloads. On the other hand, the second class of terminals can support higher bandwidth on both links and of course makes a more efficient use of the scarce frequency resource, even when nonsymmetric services are considered. The necessity to have this capability information in the network has some minor implications on control plane layers: mobility management modifications are

Table 8.9 EGPRS transmission modes.

Channel name	Code rate	Modulation	Bitrate per timeslot (kbit/s)
MCS-1	0.53	GMSK	8.8
MCS-2	0.66	GMSK	11.2
MCS-3	0.85	GMSK	14.8
MCS-4	1	GMSK	17.6
MCS-5	0.37	8-PSK	22.4
MCS-6	0.49	8-PSK	29.6
MCS-7	0.76	8-PSK	44.8
MCS-8	0.92	8-PSK	54.4
MCS-9	1	8-PSK	59.2

Table 8.10 ECSD transmission modes.

Channel name	Code rate	Modulation	Bitrate per timeslot (kbit/s)
TCH/F2.4	0.16	GMSK	3.6
TCH/F4.8	0.26	GMSK	6
TCH/F9.6	0.53	GMSK	12
TCH/F14.4	0.64	GMSK	14.5
ECSD TCS-1	0.42	8-PSK	29
ECSD TCS-2	0.46	8-PSK	32
ECSD TCS-3	0.56	8-PSK	38.8

related to the introduction of EGPRS capability information of the respective terminal. These include the multislot class as well as the EDGE modulation capabilities (downlink or uplink and downlink) and in addition an 8-PSK power class. Some modifications are also required on the RRM layer for supporting, setting up, and maintaining EGPRS temporary block flows. In addition, signaling to support the radio link control, the link quality control, and measurement procedures are introduced. There is, however, no impact on session management.

8.3.4 ECSD and EGPRS

EDGE can be used to improve both GPRS and HSCSD data services in GSM systems. In combination with EDGE, GPRS becomes EGPRS. Likewise, HSCSD becomes ECSD when enhanced by EDGE. The different achievable data rates per timeslot for the different combinations of modulation and coding schemes are summarized in Tables 8.9 and 8.10, respectively. The table shows which combinations of modulation scheme and code rate can be applied in EGPRS and ECSD, respectively. The highest data rate is achieved in EGPRS when the 8-PSK modulation scheme is combined with a code rate of one. This, however, demands extremely good channel conditions, since the code rate of one implies that there is in fact no error protection. So it will, in fact, rarely be possible to apply this mode, unless the respective application can tolerate packet losses to a good extent. The data rates that can be achieved with EGPRS and ECSD will be obviously multiples of the values in the tables when several timeslots are combined.

The link adaptation for both EGPRS and ECSD requires appropriate signaling. For this purpose, existing signaling mechanisms are applied, specifically the RR channel mode modify procedure, the assignment procedure, and the intra-cell handover procedure.

8.3.5   EDGE Classic and EDGE Compact

EDGE standardization was started when it was introduced as a work item in ETSI in 1997. The objective was to enhance the performance of HSCSD and GPRS, as described above. The result was the above-discussed so-called EDGE Phase I with ECSD and EGPRS. The EDGE specifications of Release 99 extend the previous standard by including two different air interface configurations, namely EDGE Classic and EDGE Compact. The novelty is the addition of EDGE Compact, which is only defined for packet-based services, i.e. it is only considered for EGPRS, not ECSD. The main idea behind EDGE Compact is as follows. Classical GSM systems have a relatively sparse reuse in order to secure sufficient signal quality for voice services. A typical reuse pattern in today's GSM systems would be a three sector system with a cell cluster of four. This means that the number of required channel groups is three times four, that is, 12. Given the frequency channel bandwidth of 200 kHz in GSM systems, the minimum bandwidth requirement for such a network configuration becomes 12 times 200 kHz, that is, 2.4 MHz of spectrum. For voice services, which can hardly tolerate retransmissions due to delay requirements, the C/I requirements are quite severe and cannot easily be relaxed, which means that it not advisable to work with a denser reuse. In contrast to this, data applications can tolerate delay to a greater extent. Therefore, it is possible to apply, e.g., EGPRS in a reuse scenario with a denser frequency reuse and thus a lower C/I and compensate for the higher bit error ratios induced on the channel by retransmissions and HARQ. To this end, EDGE Compact allows for a reuse configuration with three sectors and a cell cluster of just one. That means, each cell reuses the frequencies, however, within each cell three different channel groups are assigned to each of the three sectors. The minimum bandwidth requirement for such an EDGE Compact system is thus only three times 200 kHz, that is 0.6 MHz. So far it could be conceived that EDGE Compact is just EDGE Classic with a denser frequency reuse. The architectural difference, however, comes from the requirement, that in contrast to the EDGE Compact traffic channels, the control channels should still have the traditional high C/I values, known from classical GSM systems. For this purpose, GSM Compact devises a time sharing of timeslots for control channels: Co-channel sites take turns on transmitting common control channels on the downlink. The cells are separated into different time groups according to a cell cluster size of four or three, and the time group membership determines when a cell can use (downlink and uplink) control channels. For details, see, e.g., Sexton (2000) or Molkdar and Featherstone (2001).

9

Beyond GSM and UMTS: 4G

Today, many network operators have already introduced UMTS and most of them have integrated the UMTS Terrestrial Radio Access Network (UTRAN) into existing GSM backbone infrastructure and architecture. Figure 9.1 shows the respective evolution scenario for the migration from GSM to UMTS. On the basis of the existing circuit- and packet-switched infrastructure (GSM/GPRS) and entities for mobility management (HLR, MAP), the UTRAN with the new air interface has already been introduced (UMTS Phase 1). Here, UTRANs are installed in parallel to BSSs of GSM and partly even reusing existing locations. In a further step (UMTS Phase 2) the network architecture and backbone transportation facilities are adapted to the requirements of broadband packet-switched services, as the systems evolve.

In the meantime, the UMTS air interface has already been extended by the development of the fast-scheduled, packet-switched data services High Speed Downlink Packet Access (HSDPA) and High Speed Uplink Packet Access (HSUPA). Both, HSDPA and HSUPA have already been fully specified through standardization within 3GPP. Similar to EDGE in GSM, the HSxPA services apply higher-order modulation schemes and offer a series of different coding schemes, such that the data rate can be adapted to the channel conditions through the process of link adaptation. In addition, fast scheduling is applied utilizing a channel feedback with a frequency of up to once per 2 ms. Combined with clever scheduling algorithms, this is used to allow the benefits of multi-user diversity to be reaped. In theory, up to 14 Mbit/s can be achieved with HSDPA. However, in practice a couple of megabits per second are achieved. These services will further be improved, e.g., by incorporating multiple antennas at both the base station and mobile terminal and utilizing MIMO capabilities to transform the given channel diversity into link capacity gains. This is done by applying specific strategies, such as transmitting parallel data streams over different antenna pairs, and requires smart signal processing with reliable channel state information.

In parallel to the evolution of 3G systems towards UMTS/HSxPA networks, research and conceptual work is carried out towards a fully novel air interface and network architecture for future packet-switched broadband multimedia systems. Most notably the 3GPP Long Term Evolution (LTE) should be mentioned here. Even though LTE is still far from becoming an actual standard, some important properties can be listed.

GSM – Architecture, Protocols and Services Third Edition   J. Eberspächer, H.-J. Vögel, C. Bettstetter and C. Hartmann
© 2009 John Wiley & Sons, Ltd

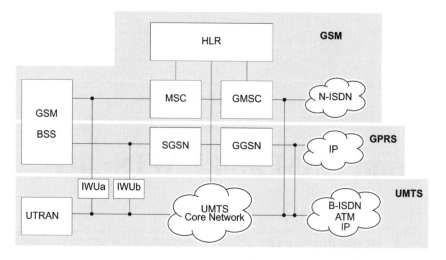

Figure 9.1 Evolution steps from GSM to UMTS.

- The air interface will be based on Orthogonal FDMA (OFDMA) on the downlink and on Single-Carrier FDMA (SC-FDMA) on the uplink.

- Fast channel feedback will allow for fast channel adaptive scheduling, achieving high data throughput.

- A frequency reuse of one is desired: it should be possible to reuse frequency resources in neighboring cells. This will, in fact, require adaptive channel and interference management strategies in order to avoid excessive interference. This is currently a major focus of research.

- The backbone node architecture will be flattened: Radio Network Controllers (RNCs), as used in UMTS in the same way as the BSC in GSM, will no longer exist. Instead, base stations will be connected directly to powerful IP routers. This implies that base stations will be much more complex than in GSM or UMTS, because functionalities such as resource allocation, packet scheduling, interference management and Hybrid ARQ (HARQ) will be moved to the base station.

At the moment it is not clear at all when a novel standard for a fourth-generation (4G) system will be available. From today's perspective it seems safe to assume that GSM networks will still be of major importance for many years. In particular, since new GSM networks will be going on air during the next couple of years in many regions of the world, the growth of GSM has not yet slowed.

Appendices

A

Data communication and networking

For the third edition of this book, we have moved this chapter on classical GSM data services from the main body to an appendix because we believe that with the advent of GSM Phase 2+ services, such as HSCSD and most prominently GPRS, the classical GSM data services are of minor importance and hardly used anymore. However, we did not want to remove them completely from the book and rather present them in this appendix for completeness.

A.1   Reference configuration

GSM was conceived in accordance with the guidelines of ISDN. Therefore, a reference configuration is also defined for GSM systems, similar to that used in ISDN systems. Using the reference configuration, one can get an impression of the range of services and the kinds of interfaces to be provided by MSs. Furthermore, the reference configuration indicates at which interface which protocols or functions terminate and where adaptation functions may have to be provided.

The GSM reference configuration comprises the functional blocks of a MS (Figure A.1) at the user–network interface Um. The mobile equipment is subdivided into a Mobile Termination (MT) and various combinations of Terminal Adapter (TA) and TE, depending on the kind of service access and interfaces offered to the subscriber.

At the interface to the mobile network, the air interface Um, MT units are defined. An integrated mobile speech or data terminal is represented only by an MT0. The MT1 unit goes one step further and offers an interface for standard-conforming equipment at the ISDN S reference point, which can be connected directly as end equipment. Likewise, normal data terminal equipment with a standard interface (e.g. V.24) can be connected via a TA and this way use the mobile transmission services. Finally, the TA functionality has been integrated into units of type MT2.

At the S or R reference point, the GSM bearer or data services are available (access points 1 and 2 in Figure A.1), whereas the teleservices are offered at the user interfaces of the TE (access point 3, Figure A.1). Among the bearer services, in addition to the transmission of digitized speech, there are circuit-switched and packet-switched data transmission. Typical

GSM – Architecture, Protocols and Services Third Edition   J. Eberspächer, H.-J. Vögel, C. Bettstetter and C. Hartmann
© 2009 John Wiley & Sons, Ltd

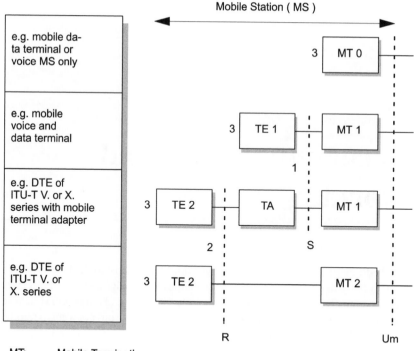

Figure A.1 GSM reference configuration.

teleservices besides telephony are, for example, SMS, group 3 fax service or emergency calls from anywhere.

A.2   Overview of data communication

The voice service needs only a switched-through physical connection, which changes its bit rate in the BSS due to the speech transcoding in the TRAU. From the MSC on, the speech signals in GSM networks are transported in standard ISDN format with a bit rate of 64 kbit/s. In comparison, realizing data services and the other teleservices such as group 3 fax is considerably more complicated. Owing to the psychoacoustic compression procedures of the GSM speech codec, data cannot be simply transmitted as a voiceband signal as in the analog network – a complete reconstruction of the data signal would not be possible. Therefore, a solution to digitize the voiceband signal similar to ISDN is not possible. Rather the available digital data must be transmitted in unchanged digital form by avoiding speech codecs in the PLMN, as is possible in ISDN. Here we have to distinguish two areas where

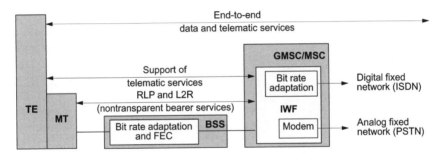

Figure A.2 Bearer services, interworking, and teleservices.

special measures have to be taken: first, the realization of data and teleservices at the air interface or within the mobile network and, second, at the transition between mobile and fixed network with the associated mapping of service features. These two areas are illustrated schematically in Figure A.2.

A PLMN offers transparent and nontransparent services. These bearer services carry data between the MT of the MS and the IWF of the MSC. For the realization of bearer services, the individual units of the GSM network define several functions:

- Bit Rate Adaptation (RA);

- Forward Error Correction (FEC);

- ARQ error correction with RLP;

- adaptation protocol L2R.

For the transmission of transparent and nontransparent data, several rate adaptation stages are required to adapt the bit rates of the bearer services to the channel data rates of the radio interface (traffic channels with 3.6, 6 and 12 kbit/s) and to the transmission rate of the fixed connections. A bearer service for data transmission can be realized in the following two ways: 9.6 kbit/s data service requires a full-rate traffic channel; all other data services can either be realized on a full-rate or half-rate channel. A MS must support both types of data traffic channels, independent of what is used for speech transmission. The data signals are transcoded first from the user data rate (9.6, 4.8, 2.4 kbit/s, etc.) to the channel data rate of the traffic channel, then further to the data rate of the fixed connection between BSS and MSC (64 kbit/s) and finally back to the user data rate. This bit rate adaptation (RA) in GSM corresponds in essence to the bit rate adaptation in the ITU-T standard V.110, which specifies the support of data terminals with an interface according to the V. series on an ISDN network (ITU-T Recommendation V.110, 2008).

On the radio channel, data are protected through the forward error correction procedures (FEC) of the GSM PLMN; and for nontransparent data services, data are additionally protected by the ARQ procedure of RLP on the whole network path between MT and MSC. Thus, RLP is terminated in the MT and MSC. The protocol adaptation to RLP of Layers 1 and 2 at the user interface is done by the L2R protocol.

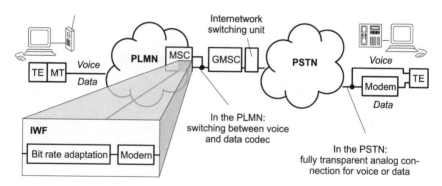

Figure A.3 Interworking scenario PLMN–PSTN for transparent data services.

Finally, the data are passed on from MSC or GMSC over an IWF to the respective data connection. The bearer services of the PLMN are transformed to the bearer services of the ISDN or another PLMN in the IWF, which is usually activated in a MSC near the MS, but could also reside in the GMSC of the network transition. In the case of ISDN this transition is relatively simple, since it may just require a potential bit rate adaptation. In the case of an analog PSTN, the available digital data must be transformed by a modem into a voiceband signal, which can then be transmitted on an analog voiceband of 3.1 kHz.

The bearer services realized in this way can offer the protocols that may be required for the support of teleservices between TE and IWF. An example is the fax adaptation protocol. The fax adapter is a special TE which maps the group 3 fax protocols with their analog physical interface upon the digital bearer services of a GSM PLMN. Thus, after another adaptation into an analog fax signal in the IWF of the MSC, it enables the end-to-end transfer of fax messages according to the ITU-T Standard T.30.

A possible interworking scenario for transparent data services of GSM with transition to a PSTN is shown in Figure A.3. The analog circuit-switched connection of the PSTN represents a transparent channel which can be used to transport arbitrary digital data signals in the voiceband. In the analog network, a subscriber selects telephone or modem depending on whether they want to transmit speech or data. In the PLMN, however, the channel coding has to be changed for different services (error protection for different bearer services; section 4.8). The bit rate adaptation has to be activated and the speech coding deactivated. In the IWF of the MSC, in addition to the bit rate adaptation, a modem needs to be added for data communication with the partner in the fixed network. In the GSM network, voice signals therefore take a different path than data signals; in the case shown in Figure A.3, the data signals are directed from the IWF to the modem, where they are digitized, then passed on after bit rate adaptation to transmission on the radio channel. In the opposite direction, the IWF passes the PCM-coded information on an ISDN channel (64 kbit/s) to the GMSC. From there it is transformed into an analog signal in a network transition switching unit and carried as a voiceband signal in the PSTN to the analog terminal.

After these introductory remarks, the GSM data and teleservices and their realization are discussed in more detail in the following sections.

A.3 Service selection at transitions between networks

A specific interworking problem arises for data services between PLMN and ISDN/ PSTN networks. Mobile-terminated calls require that the calling subscriber (ISDN or PSTN subscriber) tells the GMSC which service (speech, data, fax, etc.) they want to use. In ISDN, a Bearer Capability (BC) information element would have to be included in the SETUP message. This BC information element could then be passed on by the network transition switching unit to the GMSC and from there to the local MSC, which could thus activate the required resources. In the course of call processing (CC; section 5.4.5), the MS would also be informed about the kind of service requested by the calling subscriber and could activate the required functions. However, if there is no ISDN signaling as in analog networks, the calling subscriber is not able to do this kind of BC signaling. The service selection therefore has to use another mechanism. The GSM standard proposes two possible solutions, which are always to be used for service selection independent of the type of originating network (ISDN or PSTN).

- *Multinumbering*: the home network with this option assigns to each mobile subscriber several MSISDN numbers, each with a specific BC, which can be obtained at each call from the HLR. In this way the service that an incoming call wants is always uniquely determined. The BC information element is given to the MS when the call is being set up, so the MS can decide based on its technical features whether it wants to accept the call.

- *Single numbering*: only a single MSISDN is assigned to the mobile subscriber, and there is no BC information element transmitted with an incoming call. The MS recognizes then that a specific BC is needed when a call is accepted and requests the BC from the MSC. If the network is able to offer the requested service, the call is switched through.

Usually, the multinumbering solution is favored, since one can already verify at call arrival time in the MSC whether the requested resources are available, and the MSC side can decide about accepting the call. There is no negotiation about the BC between MS and MSC, so no radio resources are occupied unnecessarily, and the call set-up phase is not extended.

A.4 Bit rate adaptation

Five basic traffic channels are available in GSM for the realization of bearer services: TCH/H2.4, TCH/H4.8, TCH/F2.4, TCH/F4.8, TCH/F9.6 (see Tables 4.2 and 4.11) with bit rates of 3.6, 6 and 12 kbit/s. In recent standardization efforts, a TCH/F14.4 has also been defined. The bearer services (Table 4.2) with bit rates from 300 bit/s up to 9.6 kbit/s must be realized on these traffic channels. Furthermore, on the fixed connections of the GSM network, the data signals are transmitted with a data rate of 64 kbit/s.

The terminals connected at reference point R have the conventional asynchronous and synchronous interfaces. The data services at these interfaces work at bit rates as realized by GSM bearer services. Therefore, the data terminals at the R reference point have to be bit rate

Table A.1 Data rates for GSM bit rate adaptation.

Reference point	Data rate (kbit/s)		Interface (kbit/s)	
	User R	Intermediate	Radio Um	S
RA1	≤ 2.4	8		
RA1	4.8	8		
RA1	9.6	16		
RA2		8		64
RA2		16		64
RA1/RA1′		8	3.6	
RA1/RA1′		8	6	
RA1/RA1′		16	12	

adapted to the radio interface. This bit rate adaptation is derived from the V.110 standard used in ISDN in which the bit rates of the synchronous data streams are going through a two-step procedure; first, frames are formed at an intermediate rate which is a multiple of 8 kbit/s; this stream is converted to the channel bit rate of 64 kbit/s (Bocker, 1990). The asynchronous services are preprocessed by a stuffing procedure using stop bits to form a synchronous data stream.

A V.110 procedure modified according to the requirements of the air interface is also used in GSM. In essence, GSM performs a transformation of the data signals from the user data rate (e.g. 2.4 or 9.6 kbit/s) at the R reference point to the intermediate data rate (8 or 16 kbit/s) and finally to the ISDN bit rate of 64 kbit/s. The adaptation function from user to intermediate rate is called RA1; the adaptation function from intermediate rate to ISDN is called RA2. A GSM-specific bit rate adaptation step is added between the intermediate rate and the channel data rate (3.6, 6 or 12 kbit/s) of the traffic channel at the reference point Um of the air interface. This adaptation function from intermediate to channel bit rate is designated as RA1/RA1′. An adaptation function RA1′ performs the direct adaptation from user to channel data rate without going through the intermediate data rate. Table A.1 gives an overview of the bit rates at the reference points and the intermediate data rates between the RA modules.

Adaptation frames are defined for the individual bit rate adaptation steps. These frames contain signaling and synchronization data in addition to the user data. They are defined based on V.110 frames, and one distinguishes three types of GSM adaptation frames according to their length (36, 60 and 80 bits) as shown in Figures A.4 and A.5.

The conversion of data signals from user to intermediate rate in the RA1 stage uses the regular 80-bit frame of the V.110 standard. In this adaptation step, groups of 48 user data bits are supplemented with 17 fill bits and 15 signaling bits to form an 80-bit V.100 frame. Owing to the ratio 0.6 of user data to total frame length, this adaptation step converts user data rates from 4.8 into 8 kbit/s and from 9.6 into 16 kbit/s. All user data frames of less than 4.8 kbit/s are 'inflated' to a data signal of 4.8 kbit/s by repeating the individual data bits; for example, a 2.4 kbit/s signal all bits are doubled, or with a 600 bit/s signal the bits are written eight times into an RA1 frame.

Bit number

Octet number	1	2	3	4	5	6	7	8
0	0	0	0	0	0	0	0	0
1	1	Data						Signaling
2	1	Data						
3	1	Data						
4	1	Data						
5	1	Signaling						
6	1	Data						
7	1	Data						
8	1	Data						
9	1	Data						

Figure A.4 V.110 80-bit adaptation frame for the RA1 stage.

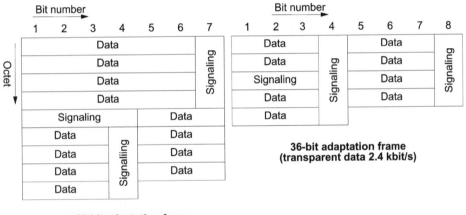

60-bit adaptation frame
(transparent data 9.6 kbit/s and 4.8 kbit/s)

36-bit adaptation frame
(transparent data 2.4 kbit/s)

Figure A.5 Modified V.110 adaptation frame for the RA1′ stage.

At the conversion of the intermediate data rate to the channel data rate in the RA1/RA1 stage, the 17 fill bits and 3 of the signaling bits are removed from the RA1 frame, since they are only used for synchronization and not needed for transmission across the air interface. This yields a modified V.100 frame of length 60 bits (Figure A.5), and the data rate is adapted from 16 to 12 kbit/s or from 8 to 6 kbit/s, respectively.

In the case of user data rates of 4.8 or 9.6 kbit/s, adaptation to the channel data rate is already complete. Only for user data rates of less than 4.8 kbit/s do additional parts of the multiple user bits need to be removed, which results in a modified V.110 frame of 36 bits. Thus, the user data rates of less than 4.8 kbit/s are adapted to a channel data rate of 3.6 kbit/s.

The user data bits of a 2.4 kbit/s signal are then no longer transmitted twice or the 600 bit/s user data signals are only written four times into the frames of the RA1′ stage. This is, however, only true for the transparent bearer services. For the nontransparent bearer services, the modified 60-bit V.110 frame is used completely for the transmission of the 60 data bits of an RLP PDU. The required signaling bits are multiplexed with user data into the RLP frame through L2R.

The modem used for communication over the PSTN resides in the IWF of the MSC, since data are transmitted from here on in digital form within the PLMN. For congestion and flow control and other functions at the modem interface, the interface signals must therefore be carried from the modem through the PLMN to the MS. For this purpose, signaling bits are reserved in the frames of the bit rate adaptation function, which represent these signals and, thus, give the MS direct modem control. The connection of such a bearer service is therefore transparent not only for user data, but also for out-of-band signaling of the (serial) modem interface in the IWF.

A.5   Asynchronous data services

Asynchronous data transmission based on the V. and X. series interfaces is widespread in fixed networks. In order to support such 'nonGSM' interfaces, the MS can include a TA over which standard terminals with a V. or X. interface (e.g. V.24) can be connected. Such an adaptation unit can also be integrated into the MS (MT2, Figure A.1).

Flow control between TA and IWF can be supported in different ways, just as in ISDN.

- *No flow control*: it is handled end-to-end in higher protocol layers (e.g. the transport layer).

- *Inband flow control*: with the X-ON/X-OFF protocol.

- *Out-of-band flow control*: according to V.110 through interface leads 105 and 106.

A.5.1   Transparent transmission in the mobile network

In the case of transparent transmission, data are transmitted with pure Layer 1 functionality. In addition to error protection at the air interface, only bit rate adaptations are performed.

User data are adapted to the traffic channel at the air interface according to the data rate and protected with forward error-correcting codes (FECs) against transmission errors. As an example, Figure A.6 shows the protocol model for transparent asynchronous data transmission over an MT1 with an S interface. Data are first converted in the TE1 or TA into a synchronous data stream by bit rate adaptation (stage RA0). In further stages, data rates are adapted with an MT1 to the standard ISDN (RA1, RA2), and then converted in MT1 over RA2, RA1 and RA1′ into the channel bit rate at the air interface. Provided with a FEC, the data are transmitted and then converted again in the BSS by the inverse operations of bit rate adaptation to 64 kbit/s at the MSC interface. However, much more frequently than an MT1 with an (internal) S interface, MSs realize a pure R interface without internal conversion to the full ISDN rate in the RA2 stage. This avoids the bit rate adaptation step RA2 and thus the conversion to the intermediate data rate in the RA1 stage. The signal is converted

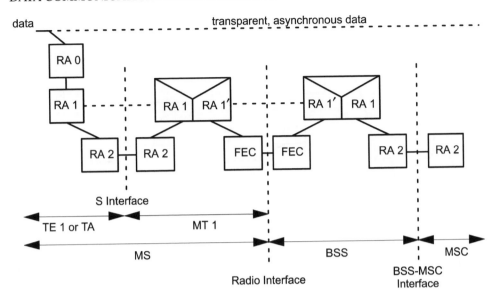

Figure A.6 Transparent transmission of asynchronous data in GSM.

immediately after the asynchronous–synchronous conversion in the RA0 stage from the user data rate to the channel data rate (stage RA1′).

A variation without a terminal adapter is shown schematically in Figure A.7. Here the complete interface functionality, Interface Circuit (I/Fcct), for a serial V. interface is integrated with the required adaptation units. The data signals D are converted into a synchronous signal in MT2 (RA0), packed into a modified V.100 frame together with signaling information S from the V. interface, and adapted to the channel data rate (RA1′). After FEC, the data signals are transmitted over the air interface and finally converted for further transmission to the data rate of an ISDN B channel after decoding and potential error correction in the BSS (RA2).

Figure A.8 shows a complete scenario with all appropriate network transitions for a transparent bearer service with modem in the interworking function for the conversion of the digital data signals into an analog voiceband signal. A mobile data terminal uses the transparent bearer service of a GSM PLMN over an R interface (it is also possible to use an S interface). The data are circuit-switched to the IWF in the MSC. To communicate with a modem in the fixed network, the IWF activates an appropriate modem function and converts the digital data signals into an analog voiceband signal. The IWF digitizes this voiceband signal again and passes the data on in PCM-coded format through the GMSC. After the network transition, the data signal is finally transmitted to the modem of the communication partner. This modem can be within a terminal in the PSTN or belong to an ISDN terminal. Before being transmitted in the PSTN, the PCM-coded signal is again converted into an analog voiceband signal. In ISDN the signal is transmitted as a PCM-coded signal of category 3.1 kHz audio; a repeated conversion is not necessary. An ISDN subscriber

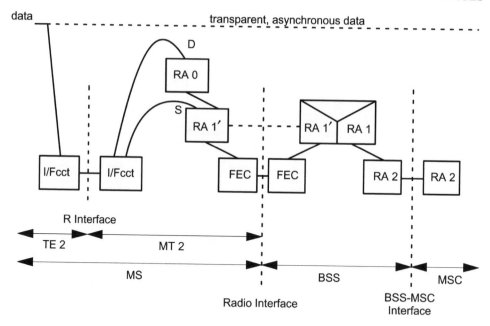

Figure A.7 Transparent transmission of asynchronous data across the R interface.

needs an adaptation unit TA′ for the conversion of the digital voiceband signal into an analog signal, which can then be processed further with a modem and passed on to the data terminal.

Another variant consists of a circuit-switched modem connection to a packet-switched network access node, as is possible from fixed connection ports. In these access nodes, the asynchronous modem signals are combined into packets in a Packet-Assembler/Disassembler (PAD) module and then transmitted through the packet-switched network. This variant of packet network access has the disadvantage that one has to switch through to the PAD over a long path, especially in the case of international roaming, since the nearest PAD is not usually the one allowed to access to packet networks.

It is also possible to connect to standard ISDN terminals without an analog modem based on the digital data transmission capability of ISDN. For this purpose, the transmission mode unrestricted digital has been defined. In this case, there is only a bit rate adaptation according to V.110 (Figure A.9). The data arrives at the MSC from the BSS in V.110 frames on an ISDN channel with 64 kbit/s and transparently is passed on to the ISDN using a B channel again in V.110 frames. The otherwise necessary modems are entirely unnecessary in the case of unrestricted digital connections. However, an ISDN subscriber can connect a terminal through an analog modem by using an adaptation unit TA′ which converts the unrestricted digital signal into a voiceband signal according to one of the V. standards.

The quality of transparent data services in GSM varies with the radio field conditions. This is illustrated by examples of comparative field measurements of the transparent data service BS26 with 9.6 kbit/s data rate, comparing a moving and a standing MS. Figure A.10 shows the weighted distribution of the bit errors of these two cases for a block length of 1024 bits. The weighted distribution indicates the frequency with which m bit errors occur in a block of

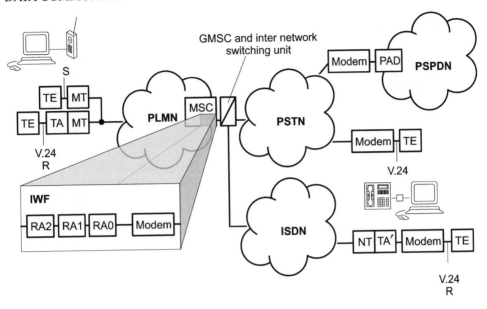

Figure A.8 Principle of transparent asynchronous data transfer (variant with a modem).

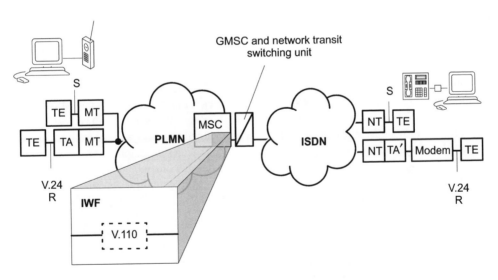

Figure A.9 Transparent data transfer to an ISDN (unrestricted digital).

P(m,1024)

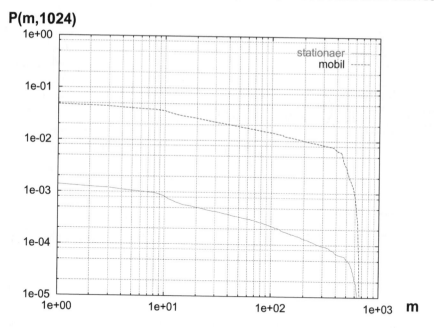

Figure A.10 Weighted distribution of a transparent GSM bearer service (BS26).

length n bits (here $n = 1024$):

$$P(m, n) = \sum_{i=m}^{\infty} P(i \text{ errors in } n\text{-bit-block}).$$

The distribution shown in Figure A.10 represents measurements of the error statistics of BS26 which were performed in 1994 in a suburban area for moving and standing MSs (Vögel *et al.*, 1995). Note that the error frequency for the standing MS ('stationaer') is clearly lower than for the moving MS ('mobil'). This result is obtained by averaging measurements over several locations and measurement tours. The resulting mean shows, for the MS, a sometimes heavily varying channel due to the unavoidable fading phenomena, which frequently cause bursts with high bit error ratios, again resulting in an aggregate higher mean bit error ratio and thus also a higher packet error ratio $P(1, 1024)$.

A.5.2 Nontransparent data transmission

In contrast to the transparent transmission mode, the nontransparent mode in GSM data services protects the user data within the PLMN through a Layer 2 protocol, the RLP, in addition to the forward error correction procedures (convolutional coding, interleaving). This protection protocol further reduces the residual bit error ratio of data transmission in the mobile network. However, the automatic repeat requests (ARQ) of the RLP introduces additional transmission delays on the data path, and the effective user data throughput is reduced (protocol overhead).

User data between MT and MSC/IWF is protected on Layer 2 by the RLP. Two kinds of transmission errors are corrected this way: first, those caused by radio interference and remaining uncorrected by forward error correction and, second, those caused through the interruptions of handover. For signaling on the FACCH, time slots are 'stolen' from the data traffic channel, which can cause data losses. The RLP also protects user data against such losses.

For nontransparent data transfer with RLP, an additional sublayer in Layer 2 is required, the L2R protocol. This relay protocol maps user data and status information of the IWF user–modem interface onto the information frames of the RLP. Depending on the kind of user data (character- or bit-oriented), one of two variations of the L2R is used: L2R Bit Oriented Protocol (L2RBOP) or L2R Character Oriented Protocol (L2RCOP). A L2R PDU is handed to RLP as an SDU and inserted into the RLP frame as a data field. The first octet of a L2R PDU always contains control information, such as the status of the signaling lines of the serial interface. Beyond that, the L2R PDU can contain an arbitrary number of such status octets. They are always inserted into the user data stream when the state of the interface changes (e.g. hardware flow control). Thus, of the 200 user data bits (Figure 5.10), only a maximum of 192 can be used for payload. However, since the signaling information is already contained in the L2R PDU, these bits must not be considered for bit rate adaptation. Thus, the modified 60-bit V.110 frame (Figure A.5) can be completely occupied with data bits, and the full channel rate of maximum 12 kbit/s can be used for the transmission of RLP data. Next come a set of four modified V.110 frames which carry a complete RLP frame. Considering the protocol overhead of RLP (16.7%) and the minimum overhead of an L2R PDU (0.5%), one obtains a usable subscriber data rate of up to 9.95 kbit/s.

The protocol model for asynchronous nontransparent character-oriented data transmission over the S interface in GSM is now presented as an example (Figure A.11). Data are transmitted by the L2R protocol, L2RCOP and RLP from the MT1 termination to the MSC. In between, as in the transparent case, are FEC, RA1, RA1' and RA2. The RLP frames are transported in a synchronous mode. Only the user data stream at the user interface is asynchronous, and in the case of the model in Figure A.11 the user data stream must be converted for the S interface into a synchronous data stream (RA0). In the case of a terminal with a V. interface and an MT2 termination (reference point R), this bit rate adaptation at the S interface would be avoided. The asynchronous data are then directly accepted at the serial interface by the L2R (I/Fcct); potential start/stop bits are removed and data are combined into L2R PDUs.

A complete scenario for network transition with nontransparent asynchronous GSM data services is shown in Figure A.12. The RLP is terminated in the MSC/IWF, and user data are converted again into an asynchronous data stream by the associated L2R. For the nontransparent case too, the IWF offers both variants for the network transition: first, using modems and, second, unrestricted digital. In the case of a network transition to the PSTN or to a 3.1 kHz audio connection into ISDN, the respective modem function is inserted and passes PCM-coded data to the GMSC (not shown in Figure A.12), which directs them into the fixed networks. Using several bit rate adaptation steps (RA0 – RA1 – RA2; Figure A.12), the user data can also be converted in the IWF into a synchronous unrestricted digital signal, which is then carried transparently over an ISDN B channel with 64 kbit/s.

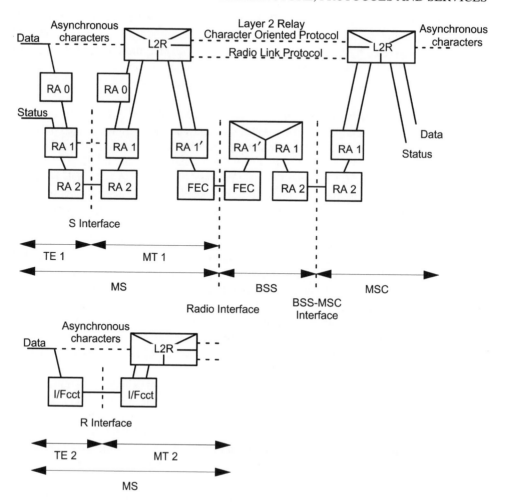

Figure A.11 Nontransparent data transmission in GSM.

A.5.3 PAD access to public packet-switched data networks

Asynchronous connection to PSPDN PADs

As shown in Figures A.8 and A.12, access to PSPDNs, e.g. Accunet in the USA or Datex-P in Germany, is already possible using the asynchronous services of GSM. This requires a PAD in the PSPDN, which packages the asynchronous data on the modem path into X.25 packets and also performs the reverse operation of unpacking. PAD access uses the protocols X.3, X.28 and X.29 (triple X profile). Just as from the fixed network, the mobile subscriber dials the extension of a PAD for access to the service of the PSPDN, provided packet network access is allowed. In this way, the subscriber has the same kind of access to the packet network as a subscriber from the fixed network, aside from the longer transmission delays

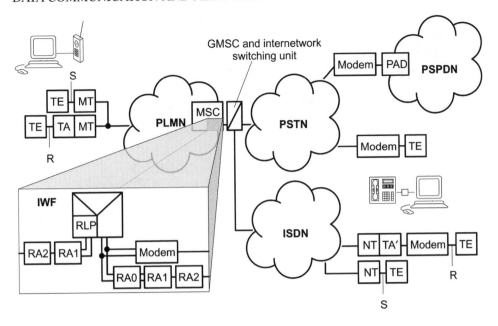

Figure A.12 Principle of nontransparent data transfer.

and the higher bit error ratios. It is therefore recommended to transmit data for PAD access across the air interface in the PLMN in nontransparent mode with RLP (Fuhrmann *et al.*, 1993).

Dedicated PAD access in GSM

However, direct access to packet data networks through the asynchronous GSM data services has disadvantages.

- One needs another subscription to a packet data network operator in addition to GSM.

- Independent of the current mobile subscriber's location, a circuit-switched connection to a PAD of a packet service provider is needed. Sometimes the packet network access is only allowed to specific PADs. This is a particular disadvantage if the mobile subscriber is currently in a foreign GSM network and incurs fees for international lines.

Therefore, GSM has defined another PSPDN access without these disadvantages: dedicated PAD access (Figure A.13). The services are defined as bearer services BS41 through BS46 (Table 4.2). With this kind of PSPDN access from a PLMN, each PLMN has at least one PAD that is responsible for the packaging/unpackaging of the X.25 packets of the respective mobile subscriber.

For this purpose, a PAD is activated as an additional resource in the IWF or in a specially reserved MSC (Figure A.13). This PAD can be reached in asynchronous mode again over transparent or nontransparent PLMN connections. However, with this solution

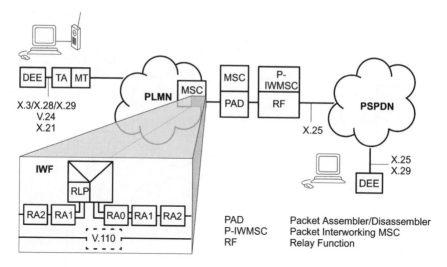

Figure A.13 Dedicated PAD access through asynchronous GSM data services.

the connection to the PAD is as short as possible, since a PAD is already reached in the nearest IWF, and international lines are never occupied for PAD access. Packetization of user data is already performed within the mobile network rather than in the remote PAD of a packet network operator, hence there is no need for a separate subscriber agreement with this packet network operator. The dedicated PAD of the current PLMN is now responsible for the packaging/unpackaging of the asynchronous data into/from X.25 packets, which are then passed on through a specific interworking MSC (P-IWMSC) to a public PSPDN (e.g. Transpac in France or Sprint's Telenet in the USA) (Fuhrmann *et al.*, 1993).

The dedicated PAD has a uniform profile in all GSM networks; it is reached in each network with the same access procedure. Even in foreign PLMNs, a MS therefore gets the earliest and lowest-cost access to the packet data network. Charging occurs to the account of the GSM extension (MSISDN) of the mobile subscriber; a separate Network User Identification (NUI) for the PSPDN is not necessary. However, only outgoing packet connections are possible.

A.6   Synchronous data services

A.6.1   Overview

Synchronous data services allow access to synchronous modems in the PSTN or ISDN as well as to circuit-switched data networks. Such access is not very significant; however, synchronous data services are defined in GSM. The essential differences to the asynchronous data transmission procedures are in bit rate adaptation and modems. For a synchronous data service, no RA0 bit rate adaptation is needed (conversion from asynchronous to synchronous), since data are already in synchronous format. Instead, special synchronous modems are required in the IWF. Synchronous data services can only be offered in

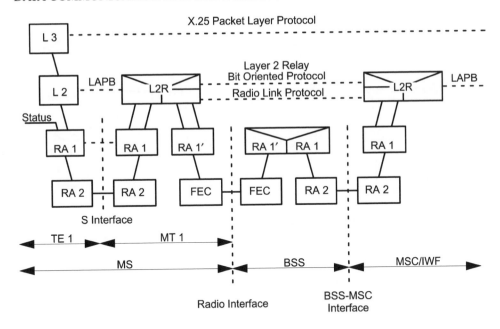

Figure A.14 X.25 access at the ISDN S interface.

transparent mode, with the exception of access to X.25 packet networks, which are a significant application of synchronous data service in GSM.

A.6.2 Synchronous X.25 packet data network access

The protocol model shown in Figure A.14 is the model for synchronous data transmission in nontransparent mode with the packet data access protocol according to the ITU-T standard X.25. Owing to the nontransparent transmission procedure, the X.25 Link Access Procedure B (LAPB) must be terminated in the MT as well as in the IWF. Since LAPB of the X.25 protocol stack operates in a bit-oriented mode, the L2RBOP is required.

Basic packet mode

Two variants of PSPDN access can be realized with the protocol model in Figure A.14:

- PSPDN access according to ITU-T X.32;

- access to PSPDN packet handlers according to ITU-T X.31 Case A (basic packet mode); PSPDN access according to X.32 has not met much acceptance with the applications; the X.31 procedure is used more often.

PSPDN access according to X.32 is the simpler variant. The X.25 packets can be transferred directly over a synchronous modem in the IWF to the PSPDN. This does not necessarily require nontransparent transmission in GSM, but it helps because of the lower bit error ratio. In case of the nontransparent transmission, the LAPB protocol has to be

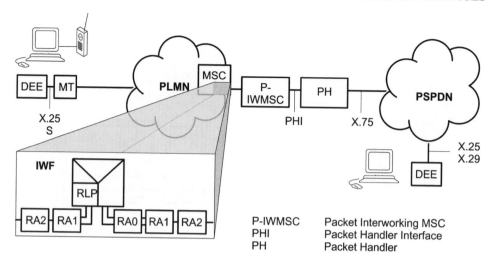

Figure A.15 Dedicated packet mode with packet handler in GSM.

terminated in MT and IWF (see above). The subscriber needs a NUI. Incoming and outgoing packet connections are possible, but again there is the problem of needing circuit-switched connections to the home PLMN, just as in the case of international roaming.

The access procedure according to ITU-T standard X.31 Case A (basic packet mode) is the more favored variant of this group of services. The X.25 packets of the mobile subscriber are passed from the IWF to the packet handler of the ISDN. Since speed adaptation with the X.31 procedure is performed in the ISDN B channel by flag stuffing, the protocol has to be terminated in the IWF, which means that the nontransparent mode of GSM can be used. In this case too, there are connections possible only through the packet handler of the home network.

Dedicated packet mode

As in the case of asynchronous PAD access to packet data networks (see section A.5.3), the synchronous case of the X.25 access protocol also offers an alternative, which allows the most immediate transition to the PSPDN, even if the MS is in a foreign network (international roaming). For this purpose, a dedicated mode is also defined with each PLMN having its own packet handler.

Figure A.15 shows the principle of this dedicated packet mode. It essentially includes the functions of the basic packet mode, with the difference that the packet handler is integrated into the GSM network. Data are transmitted in the PLMN in nontransparent and synchronous mode. Access to the packet handler is the same in all PLMNs, and is also available to foreign mobile subscribers. Since packet data networks do not know roaming, only outgoing data calls are possible.

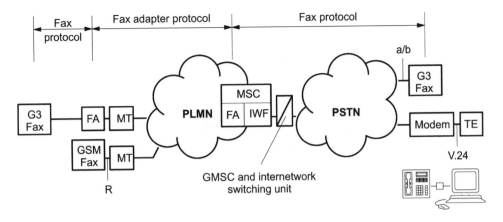

Figure A.16 Fax adapter in GSM.

A.7 Teleservices: fax

In the following, we briefly explain the realization of the GSM fax service.

The GSM standard considers the connection of a regular group 3 fax terminal with its two-wire interface to an appropriately equipped MS as the standard configuration of a mobile fax application. The GSM fax service is supposed to enable this configuration to conduct fax transmissions with standard group 3 fax terminals over mobile connections. This requires mapping the fax protocol of the analog two-wire interface onto the digital GSM transmission, because the fax procedure defines a complete protocol stack with its own modulation, coding, user data compression, inband signaling, etc. Therefore, the Fax Adapter (FA) has been defined for this mapping. The principle is summarized in Figure A.16. The FA of the MS converts the fax protocol of a standard group 3 fax terminal on an analog two-wire line into a GSM internal FA protocol. The PDUs of the adapter protocol are transmitted over the MT with the GSM data services to the FA in the IWF of the MSC, and there they are again converted into the T.30 protocol on the analog line, or transmitted to the ISDN in PCM-coded format, where again a terminal adapter allows the connection of a group 3 fax terminal (not shown in Figure A.16).

A more compact version of a mobile fax terminal is possible if the FA and the group 3 fax terminal are integrated into a compact terminal (GSM fax, Figure A.16). Such equipment can be connected at reference point R to an MT2. It delivers a digital signal directly. With this integration, the analog two-wire line interface becomes superfluous.

Hence, all of the analog functions can be omitted such as demodulation and digitalization of the group 3 fax signals, i.e. the GSM fax needs no analog components such as a modem building block. Therefore, the integrated GSM fax must implement the FA protocol and terminate it at reference point R in order to guarantee correct control of the analog fax components in the FA of the IWF.

A complete fax scenario with the required analog components in the FA is illustrated in Figure A.17. The FA needs several function blocks in the MS as well as in the IWF for the conversion of the fax protocol of the analog a/b interface to the digital transmission procedure

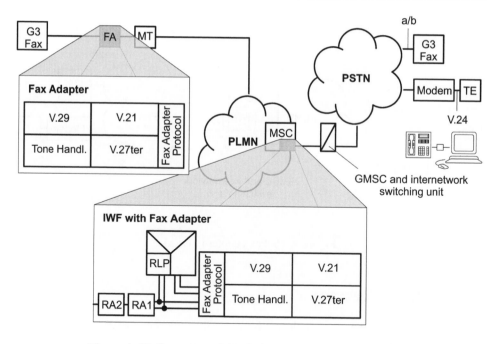

Figure A.17 Overview of the GSM procedure for fax service.

in the PLMN. Group 3 fax equipment according to ITU-T standard T.30 employs three modem building blocks, all three operating in half-duplex mode. A V.21 modem (300 bit/s) is used for the signaling phase to set up a fax connection, whereas the information transfer phase uses a V.27ter modem (4.8 or 2.4 kbit/s) or a V.29 modem (9.6, 4.8 and 2.4 kbit/s). For conversion from analog signaling tones into messages of the fax adapter protocol, the fax adapter needs an additional tone handler. It is used to transfer the (re)digitized fax signals over either a transparent or a nontransparent GSM bearer service to the IWF in the MSC. The fax adapter protocol provides a complete mapping of the T.30 protocol, such that the IWF is able to handle the complete fax protocol with the partner entity. From the view of the partner entity, the complete GSM connection consisting of FA, MS and IWF represents a physical connection with a mobile group 3 fax terminal.

Fax service poses special demands on the service quality of the data channel which the GSM network provides for this teleservice. In particular, propagation delays must remain below a maximum threshold, since timers of the T.30 protocol expire otherwise. This is especially critical if RLP is used to reduce transmission errors, which introduces additional delays into the data path through its ARQ procedure.

Two fax services are specified: the transparent procedure and the nontransparent procedure, depending on the kind of bearer service used. The resulting protocol models are shown in Figures A.18 and A.19, respectively. It is evident that in both cases the fax protocol is superimposed onto the respective bearer service. The transparent fax procedure is easier to realize, although with its transparent bearer service, it incurs a correspondingly variable fax quality.

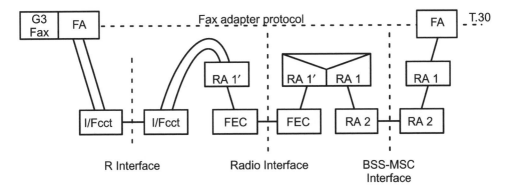

Figure A.18 Transparent fax procedure in GSM radio interface BSS–MSC.

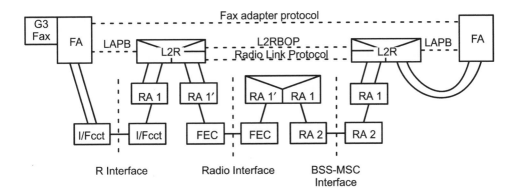

Figure A.19 Nontransparent fax procedure in GSM.

In contrast, the nontransparent fax procedure based on RLP is very well protected against transmission errors, and it delivers very acceptable quality of the transmitted documents over a wide range of distances. However, due to the varying transmission conditions, there are also variable delays of the RLP which lead to intermediate buffering of fax signals in the FA and which, in the worst case, can cause the breakdown of the fax transfer (Decker and Pertz, 1993).

B

Aspects of network operation

For the efficient and successful operation of a modern communication network such as a GSM PLMN, comprehensive Network Management (NM) is mandatory. NM encompasses all functions and activities which control, monitor and record the usage and resource performance of a telecommunication network, with the objective of offering the subscribers telecommunication services of a certain objective level of quality. Various aspects of quality are either defined and prescribed in standards or laid down in operator-specific definitions. Special attention has to be paid to the gap between (mostly simple) measurable technical performance data of the network and the QoS experienced (subjectively) by the subscriber. Modern NM systems should therefore also include (automated) capabilities to accept reports and complaints from subscribers and convert them into measures to be taken by NM (e.g. trouble ticketing systems).

B.1 Objectives of GSM NM

Along with the communication network which realizes the services with its functional units (MS, BSS, MSC, HLR, VLR), one needs to operate a corresponding NM system for support and administration. This NM system is responsible for operation and maintenance of the functional PLMN units and the collection of operational data. The operational data comprise all of the measurement data which characterize performance, load, reliability and usage of the network elements, including times of usage by individual subscribers, which are the basis for the calculation of connection fees (billing). Furthermore, in GSM systems in particular, the techniques supporting security must have counterparts in security management functions of network management. This security management is based on two registers: the Authentication Center (AUC) provides key management for authentication and encryption and the Equipment Identity Register (EIR) provides barring of service access for individual equipment, 'blacklisting'. To summarize, for all functions of the telecommunication network and its individual functional units (network elements), there are corresponding NM functions. The GSM standard has defined the following overall objectives of NM:

GSM – Architecture, Protocols and Services Third Edition   J. Eberspächer, H.-J. Vögel, C. Bettstetter and C. Hartmann
© 2009 John Wiley & Sons, Ltd

- international operation of NM;

- cost limitation of GSM systems with regard to short-term as well as long-term aspects;

- achievement of service quality which at least matches the competing analog mobile radio systems.

The international operation of a GSM system includes among others the interoperability with other GSM networks (including different countries) and with ISDN networks, as well as the information exchange among network operators (billing, statistical data, subscriber complaints, invalid IMEI, etc.). These NM functions are in large part necessary for network operation allowing international roaming of subscribers, and therefore they must be standardized. Their implementation is mandatory.

The costs of a telecommunication system consist of invested capital and operational costs. The investments comprise the cost of the installation of the network and of the NM, as well as development and licensing costs. The periodically incurred costs include operation, maintenance and administration as well as interest, amortization and taxes. Lost revenues due to failing equipment or partial or complete network failure must be included in the periodically incurred costs, whereas consequential losses due to cases of failure, e.g. because of lost customers, cannot be estimated and included. Therefore, the reliability and maintainability of the network equipment is of course of immense importance and has a large impact on costs. The installation of a NM system on the one hand increases the need for investment capital for the infrastructure as well as for spare capacities in the network. On the other hand, these costs for a standardized comprehensive NM system must be compared with the expenses for administration, operation and maintenance of network elements with manufacturer-proprietary management, or the costs which arise from not recognizing and repairing network failures early enough. Therefore, it has to be the objective of a cost-efficient NM system to define and implement uniform vendor-independent NM concepts and protocols for all network elements, and also to guarantee interoperability of network components from different manufacturers through uniform interfaces within the network.

The QoS to be achieved can be characterized with technical criteria such as speech quality, bit error ratio, network capacity, blocking probability, call disconnection rates, supply probability and availability, and it can also be characterized with nontechnical criteria such as ease of operation and comfort of subscriber access or even hotline and support services.

Considering these objectives, the following functional areas for NM systems can be identified:

- administrative and business area (subscribers, terminal equipment, charging, billing, statistics);

- security management;

- operation and performance management;

- system version control;

- maintenance.

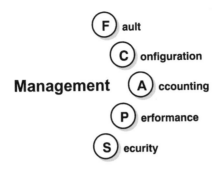

Figure B.1 Functional areas of TMN systems.

These functions are realized in GSM based on the concept of the TMN. In general, they are summarized with the acronym FCAPS – Fault, Configuration, Accounting, Performance and Security management (Figure B.1).

Fault management includes functions such as failure recognition, failure diagnosis, alarm administration and filtering as well as capabilities for the identification of causes of failures or alarms and keeping of failure logs. Configuration management administers network configurations and handles changes, activates/deactivates equipment and provides tools for the automatic determination of network topology and connectivity. Accounting management deals with the subscribers and is responsible for the establishment and administration of subscriber accounts and service profiles. Periodic billing for the individual subscribers originates here, based on measured usage times and durations; statistics are calculated, in certain circumstances only for network subareas (billing domains). In performance management, one observes, measures and monitors performance (throughput, failure rates, response times, etc.) and utilization of network components (hardware and software). The objective is on the one hand to ensure a good utilization of resources and on the other hand to recognize trends leading to overload and to be able to start counter-measures early enough. Finally, security management provides thorough access control, the authentication of subscribers and an effective encryption of sensitive data.

B.2 Telecommunication management network

TMN was standardized within ITU-T/ETSI/CEPT almost simultaneously with the pan-European mobile radio system GSM. The guidelines of the M. series of the ITU-T (M.20, M.30) serve as a framework.

TMN defines an open system with standardized interfaces. This standardization enables a platform-independent multivendor environment for management of all components of a telecommunication network. Essentially, it realizes the communication of a management system with network elements it administers, which are considered as managed objects. These objects are abstract information models of the physical resources. A manager can send commands to these administered objects over a standardized interface, can request or change parameters, or be informed by the objects about events that occurred (notification). For this purpose, an agent resides in the managed object, which generates the management

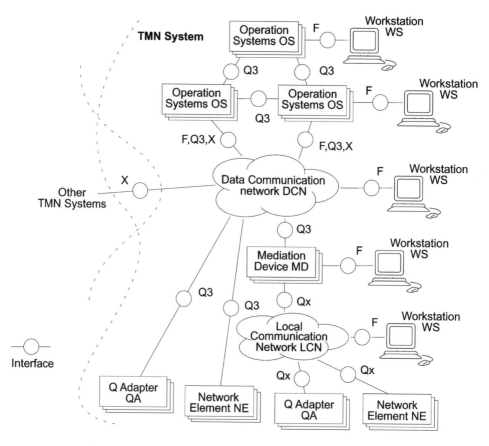

Figure B.2   TMN architecture (according to M.3010 (CCITT Recommendation M.3010, 1992)).

messages or evaluates the requests from the manager, and maps them onto corresponding operations or manipulations of the physical resources. This mapping is system specific as well as implementation dependent and, hence, not standardized. The generalized architecture of a TMN is illustrated in Figure B.2.

The NM proper is realized in an Operation System (OS). The OSs represent the surveillance and control systems of a TMN system. These systems can communicate with each other directly or form hierarchies. A standardized interface Q3 serves for the communication of the OSs within a TMN, whereas the interconnection of two TMN systems occurs over the X interface (Figure B.2). The management functionality can also be subdivided into several logical layers according to the OSI hierarchy. For this approach, TMN provides the Logical Layered Architecture (LLA) as a framework. The exact numbering and corresponding functionality of each LLA plane were not yet finalized in the standardization process at the time of writing, however, the following planes have been found to be useful (Figure B.3): Business Management Layer (BML), Service Management Layer (SML), Network Management Layer (NML) and Element Management Layer (EML).

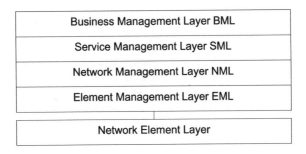

Figure B.3 Logical layered architecture of a TMN system.

The TMN functions of the EML are realized by the network elements (NEs) and contain basic TMN functions such as performance data collection, alarm generation and collection, self diagnosis, address conversion and protocol conversion. Frequently, the EML is also known as a Network EML (NEML) or a Subnetwork Management Layer (SNML) (Sahin, 1993). The NML-TMN functions are normally performed by operation systems and used for the realization of NM applications, which require a network-wide scope. For this purpose, the NML receives aggregate data from the EML and generates a global system view from them. On the SML plane, management activities are performed which concern the subscriber and their service profile rather than physical network components. The customer contact is administered in the SML, which includes functions such as establishing a subscriber account, initializing supplementary services and several others. The highest degree of abstraction is reached in the BML, which has the responsibility for the total network operation. The BML supports strategic network planning and the cooperation among network operators (Sahin, 1993).

For example, an operation system OS could act as a basic OS and be in charge only of a region with a subset of NEs, or it could be a Network OS which communicates with all of the basic OSs and implements network-wide management functionality. As a service OS, an OS assumes network-wide responsibility for the management of one service, whereas on the BML plane, care is taken for charging, billing and administration of the whole network and its services.

The individual functional units of the telecommunication network are mapped into NEs. These elements are abstract representations of the physical components of the telecommunication network, which is administered by this TMN. The OSs communicate with the network elements over a comprehensive data network, a Data Communication Network (DCN). For this purpose, an interface Q3 has been defined, whose protocols comprise all seven layers of the OSI model. However, not every NE must support the full range of Q3 interface capabilities.

For NEs whose TMN interface contains a reduced range of functionality (Qx), a Mediation Device (MD) is interposed, which essentially performs the task of protocol conversion between Qx and Q3. A mediator can serve several NEs with incomplete Q3 interfaces, which can be connected to the mediator through a Local Communication Network (LCN). The functions of a mediator are difficult to define in general and depend on the respective application, since the range of restrictions of a Qx interface with regard to the Q3 interface is not standardized (Glitho and Hayes, 1995). Therefore, a mediator could, for example, realize functions such as data storage, filtering, protocol adaptation or data aggregation and compression.

In spite of ongoing TMN standardization, new network elements and systems without a TMN interface are continuously added and must be integrated. For such cases, the function of the Q adapter (QA) has been defined. In contrast to the mediator MD, which is prefixed to TMN-capable devices with reduced functionality at the Q interface, a QA allows integration of devices which are not TMN capable, and the QA must therefore be tailored for each respective device.

Finally, the operator personnel have access to the TMN system at the F interface through management Workstations (WSs) in order to perform management transactions and to check or change parameters. Thus, a TMN system gives the network operator at a WS the capability to supply any network element with configuration data, to receive and analyze failure reports and alarms or to download locally collected measurement data and usage information. The TMN protocol stack required for this purpose is based on OSI protocols and comprises all seven layers (see also Figure B.6). The main element of the TMN protocol architecture is the Common Management Information Service Element (CMISE) from the OSI system management, which resides in the application layer (OSI Layer 7) (Sahin, 1993). The CMISE consists of a service definition, the Common Management Information Service (CMIS), and a protocol definition, the Common Management Information Protocol (CMIP). The CMISE defines a uniform message format for requests and notifications between management OS and the managed elements NE or the respective QA.

B.3   TMN realization in GSM networks

TMN and GSM were standardized at approximately the same time, so that there was a good opportunity to apply TMN principles and methods in a complete TMN system for NM in GSM from the beginning and from ground up. For this purpose, specific working groups were founded for the five TMN categories (Figure B.1) as well as for architecture and protocol questions which were supposed to develop as much as possible of the TMN system and its services, while following the top-down methodology (CCITT Recommendation M.3010, 1992; CCITT Recommendation M.3020, 1992) recommended by the ITU-T. This objective could be pretty much achieved, only that the development methodology was complemented by a bottom-up approach which was rooted in the detailed knowledge about the network components being specified at the same time. The intent was to reach the objective of a complete standard earlier (Lin, 1997; Towle, 1995). The five TMN categories are essentially realized for all of the GSM system; however, there are some limitations in failure, configuration and security management. Failure and configuration management are specified only for the BSS; the reasons are that on the one hand the databases (HLR, VLR) were assigned to accounting management, and on the other hand standardization efforts were to concentrate on GSM-specific areas. Concentration on GSM-specific areas thereby excluded failure and configuration management for the MSC, which from the management point of view is essentially a standard ISDN switching exchange. For the same reasons, security management is also limited to GSM-specific areas.

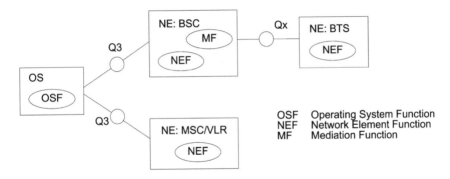

Figure B.4 A simple TMN architecture of a GSM system (according to Towle (1995)).

The resulting GSM TMN architecture is shown schematically in Figure B.4. In GSM, the BSC and the MSC have a Q3 interface as network elements to the OS. In addition to the BSS management, the BSC NE always contains a Mediation Function (MF) and a Qx interface to the NE supporting the BTS functionality.

An object-oriented information model of the network has been defined for the realization of the GSM TMN services. The model contains more than 100 Managed Object Classes (MOCs) with a total of about 500 attributes. This includes the ITU-T standard objects as well as GSM-specific objects, which include the GSM NEs (BSS, HLR, VLR, MSC, AUC, EIR) on the one hand, but also represent network and management resources (e.g. for SMS service realization or for file transfer between OS and NE) as managed objects. These objects usually contain a state space and attributes which can be checked or changed (request) as well as mechanisms for notification, which report the state or attribute changes. In addition, there are commands for creation or deletion of objects, e.g. in the HLR with create/modify/delete subscriber or create/modify/delete MSISDN or in the EIR with create/interrogate/delete equipment (Towle, 1995). File transfer objects are used, in particular, in the information model dealing with the registers, since it involves the movement of large amounts of data.

The TMN communication platform to be used as a DCN can be either an OSI X.25 packet network or the SS#7 signaling network (MTP and SCCP). Both offer a packet-switching service which can be used to transport management messages. Each NE is connected to this MN over a Management Network Access Point (MNAP); see Figure B.5.

If the TMN uses X.25, the DCN can be the public PSPDN or a dedicated packet-switching network within the PLMN with the MSC as a packet-switching node. In addition, the MSC can include an interworking function for protocol conversion from an external X.25 link to the SS#7 SCCP, which realizes the connection of the OMC to the PLMN through an external X.25 link. Further transport of management messages is then performed by the SS#7 network internal to the PLMN.

The framework defined for the GSM TMN protocol stack at the Q3 interface is presented in Figure B.6. The end-to-end transport of messages between OS and NE is realized with the OSI Class 2 transport protocol (TP2), which allows the setup and multiplexing of end-to-end transport connections over an X.25 or SCCP connection. Error detection and data security are

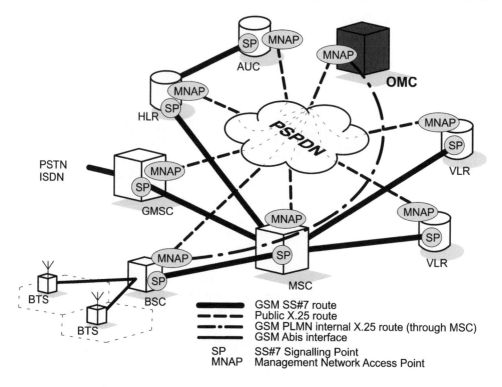

Figure B.5 Potential signaling interfaces in a GSM TMN.

not provided in TP2; they are not needed since X.25 as well as SCCP offer a secure message transport service already.

Of course, the OSI protocol stack also needs the protocols for the data link and presentation layers. The OSI CMISE plays the central role in GSM network management. Its services are used by a System Management Application Process (SMAP) to issue commands, to receive notifications, to check parameters, etc. For file transfer between objects, GSM TMN uses the OSI File Transfer Access and Management (FTAM) protocol. It is designed for the efficient transport of large volumes of data.

CMISE needs a few more Service Elements (SEs) in the application layer for providing services: the Association Control Service Element (ACSE) and the Remote Operations Service Element (ROSE). The ACSE is a sublayer of the application layer which allows application elements (here CMISE) to set up and take down connections between each other. The ROSE services are realized with a protocol which enables initiation or execution of operations on remote systems. In this way ROSE implements the paradigm also known as Remote Procedure Call (RPC).

There is also a management system for the signaling components of a GSM system. This SS#7 SMAP uses the services of the Operation Maintenance and Administration Part (OMAP) which allows observation, configuration and control of the SS#7 network resources. Essentially, the OMAP consists of two Application Service Elements (ASEs), the MTP

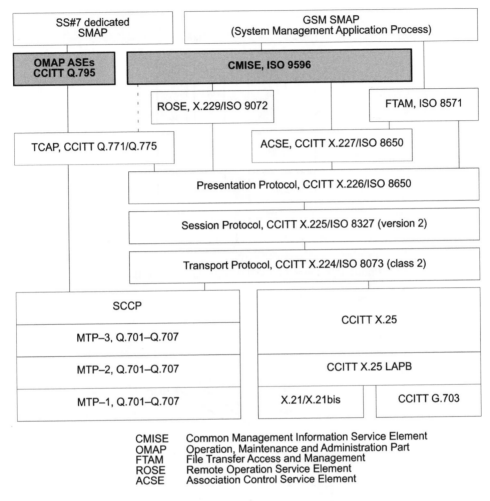

CMISE	Common Management Information Service Element
OMAP	Operation, Maintenance and Administration Part
FTAM	File Transfer Access and Management
ROSE	Remote Operation Service Element
ACSE	Association Control Service Element

Figure B.6 GSM network management protocols at the Q3-interface.

Routing Verification Test (MRVT) and the SCCP Routing Verification Test (SRVT), which allow verification of whether the SS#7 network works properly on the MTP or SCCP planes. Another management application part is BSSOMAP which is used to transport management messages from OMC to BSC through the MSC over the A interface and to execute management activities for the BSS (Figure B.7; and compare it with Figure 5.11) (Schmidt, 1993).

Network management is usually organized in a geographically centralized way. For the remote surveillance and control of NM functions there are usually one or more Operation and Maintenance Centers (OMCs). For efficient NM, these OMCs can be operated as regional subcenters according to the LLA hierarchy of the various TMN management planes, and they can be combined under a central Network Management Center (NMC); see Figure B.8.

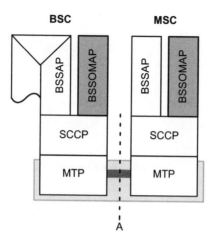

Figure B.7 Operation and maintenance of the BSS.

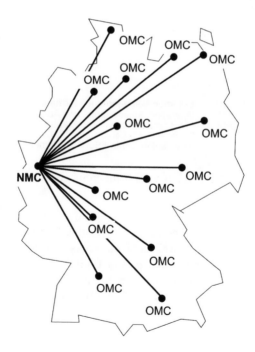

Figure B.8 Hierarchical organization of network management within Germany.

C

GSM Addresses

BCC Base Transceiver Station Color Code

BSIC Base Transceiver Station Identity Code

(M)CC (Mobile) Country Code

CI Cell Identifier

FAC Final Assembly Code

GCI Global Cell Identity

IMEI International Mobile Equipment Identity

IMSI International Mobile Subscriber Identity

LAC Location Area Code

LAI Location Area Identity

LMSI Local Mobile Subscriber Identity

MNC Mobile Network Code

MSIN Mobile Subscriber Identification Number

MSISDN Mobile Station ISDN Number

MSRN Mobile Station Roaming Number

NCC Network Colour Code

NDC National Destination Code

NMSI National Mobile Subscriber Identity

SN Subscriber Number

SNR Serial Number

TAC Type Approval Code

TMSI Temporary Mobile Subscriber Identity

Additional Addresses for GPRS

APN Access Point Name

GSN Address GPRS Support Node Address (e.g. IP address of SGSN and GGSN)

GSN Number GPRS Support Node Number (for SGSN and GGSN for communication with e.g. HLR and VLR)

IP Address Internet Protocol address

NSAPI Network layer Service Access Point Identifier

P-TMSI Packet Temporary Mobile Subscriber Identity

PDP Address PDP Address (e.g. IP address, X.25 address)

RAC Routing Area Code

RAI Routing Area Identity

TID Tunnel Identifier (= MSI + NSAPI)

TLLI Temporary Logical Link Identifier

D

List of Acronyms

3GPP 3rd Generation Partnership Project

3GPP2 3rd Generation Partnership Project 2

3PTY Three Party Service

8-PSK 8 Phase Shift Keying

A3, A5, A8 Encryption Algorithms

AB Access Burst

Abis BTS-BSC Interface

ACELP Algebraic Code Excitation Linear Prediction

ACSE Association Control Service Element

AGCH Access Grant Channel

AMR Adaptive Multi-rate (codec)

AOC Advice of Charge

ARQ Automatic Repeat Request

ASCI Advanced Speech Call Items

ASE Application Service Element

ATM Asynchronous Transfer Mode

AUC Authentication Center

BAIC Barring of All Incoming Calls

BAOC Barring of All Outgoing Calls

BCC Base Transceiver Station Color Code

BCCH Broadcast Control Channel

BCH Broadcast Channel

BCS Block Check Sequence

BFI Bad Frame Indication

BG Border Gateway

BIC-Roam Barring of Incoming Calls when Roaming Outside the Home PLMN

B-ISDN Broadband ISDN

Bm Mobile B Channel

BN Bit Number

BOIC Barring of Outgoing International Calls

BOIC-exHC Barring of Outgoing International Calls except those to Home PLMN

BSC Base Station Controller

BSIC Base Tranceiver Station Identity Code

BSN Block Sequence Number

BSS Base Station Subsystem

BSSAP Base Station System Application Part

BSSAP+ Base Station System Application Part +

BSSGP Base Station System GPRS Application Protocol

BSSMAP Base Station System Management Application Part

BTS Base Transceiver Station

BTSM Base Transceiver Station Management

CA Cell Allocation

CAMEL Customized Applications for Mobile Network Enhanced Logic

CAP CAMEL Application Part

CBCH Cell Broadcast Channel

CC Country Code

CCBS Completion of Call to Busy Subscriber

CCCH Common Control Channel

CDMA Code Division Multiple Access

CELP Code Excited Linear Prediction Coding

CFB Call Forwarding on Mobile Subscriber Busy

CFNRc Call Forwarding on Mobile Subscriber Not Reachable

CFNRy Call Forwarding on No Reply

CFU Call Forwarding Unconditional

CI Cell Identifier

C/I Carrier-to-Interference Ratio

CIR Carrier-to-Interference Ratio

CLIP Calling Number Identification Presentation

CLIR Calling Number Identification Restriction

CLNS Connectionless Network Service

CM Connection Management

CMISE Common Management Information Service Element

CN Core Network

CODEC Coder/Decoder

COLP Connected Line Identification Presentation

COLR Connected Line Identification Restriction

CONF Conference Calling

CONS Connection-Oriented Network Service

CPS Coding and Puncturing Scheme

CRC Cyclic Redundancy Check

CT Call Transfer

CUG Closed User Group

CW Call Waiting

DAB Digital Audio Broadcast

DB Dummy Burst

DCCH Dedicated Control Channel

DCN Data Communication Network

DECT Digital Enhanced Cordless Telecommunication

DHCP Dynamic Host Configuration Protocol

Dm Mobile D Channel

DNS Domain Name Service

DRX Discontinuous Reception

DSL Digital Subscriber Line

DTMF Dual Tone Multiple Frequency

DTX Discontinuous Transmission

DTAP Direct Transfer Application Part

DVB Digital Video Broadcast

ECSD Enhanced Circuit Switched Data

EDGE Enhanced Data Rates for Global (GSM) Evolution

EFR Enhanced Full Rate (CODEC)

EGPRS Enhanced GPRS

EIR Equipment Identity Register

EMLPP Enhanced Multi-Level Precedence and Pre-emption Service

EMS Enhanced Messaging Service

E-OTD Enhanced Observed Time Difference

ERMES European Radio Messaging Standard

ETSI European Telecommunication Standards Institute

FA Fax Adapter

FAC Final Assembly Code

FACCH Fast Associated Control Channel

FCAPS Fault, Configuration, Accounting, Performance, Security

FB Frequency Correction Burst

FCCH Frequency Correction Channel

FCS Frame Check Sequence

FDD Frequency Division Duplex

FDMA Frequency Division Multiple Access

FEC Forward Error Correction

FN TDMA Frame Number

FPH Freephone Service

FPLMTS Future Public Land Mobile Telecommunication System

FTAM File Transfer Access and Management

GCR Group Call Register

GEA GPRS Encryption Algorithm

GGSN Gateway GPRS Support Node

GMLC Gateway Mobile Location Center

GMM/SM GPRS Mobility Management and Session Management protocol

GMSC Gateway MSC

GMSK Gaussian Minimum Shift Keying

GPRS General Packet Radio Service

GPS Global Positioning System

GSC GSM Speech Codec

GSM Global System for Mobile Communication

GSMSS GSM Satellite System

GSN GPRS Support Node

GTP GPRS Tunnelling Protocol

HARQ Hybrid Automatic Repeat Request

HCS Header Check Sequence

HLR Home Location Register

HSCSD High Speed Circuit Switched Data

HSDPA High Speed Downlink Packet Access

HSN Hopping Sequence Number

HSUPA High Speed Uplink Packet Access

HTML Hypertext Markup Language

HTTP Hypertext Transport Protocol

IMEI International Mobile Equipment Identity

IMSI International Mobile Subscriber Identity

IMT-2000 International Mobile Telephone System 2000

IN Intelligent Network

INAP Intelligent Network Application Part

IP Internet Protocol

IPv4 Internet Protocol Version 4

IPv6 Internet Protocol Version 6

ISC International Switching Center

ISDN Integrated Services Digital Network

IWF Interworking Function

Kc Cipher/Decipher Key

Ki Subscriber Authentication Key

L2R Layer 2 Relay

L2RBOP Layer 2 Relay Bit-Oriented Protocol

L2RCOP Layer 2 Relay Character-Oriented Protocol

LA Location Area

LAC Location Area Code

LAI Location Area ID

LAPDm Link Access Procedure D mobile

LCN Local Communication Network

LCS Location Service

LEO Low Earth Orbiting satellite

LLC Logical Link Control layer

LMSI Local Mobile Subscriber Identity

LPC Linear Predictive Coding

LTE Long Term Evolution

LTP Long Term Prediction

MA Mobile Allocation

MAC Medium Access Control layer

MAH Mobile Access Hunting

MAIO Mobile Allocation Index Offset

MAP Mobile Application Part

MC Multi Carrier

MCI Malicious Call Identification

MD Mediation Device

MEO Medium Earth Orbiting satellite

MexE Mobile Station Application Execution Environment

MHS Message Handling System

MIMO Multiple Input Multiple Output

MM Mobility Management

MMI Man–Machine Interface

MMS Multimedia Messaging Service

MMS-C Multimedia Messaging Service-Center

MMSE Multimedia Messaging Service Environment

MMSNA Multimedia Messaging Service Network Architecture

MNC Mobile Network Code

MOS Mean Opinion Score

MS Mobile Station

MSC Mobile Switching Center

MSIN Mobile Subscriber Identification Number

MSISDN Mobile Station ISDN Number

MSK Minimum Shift Keying

MSRN Mobile Station Roaming Number

MSS Mobile Satellite System

MT Mobile Termination

MTP Message Transfer Part

NB Normal Burst

NCC Network Colour Code

NCH Notification Channel

NDC National Destination Code

NE Network Element

NMSI National Mobile Subscriber Identity

NMT Nordic Mobile Telephone

NS Network Service

NSAPI Network Layer Service Access Point Identifier

OFDMA Orthogonal Frequency Division Multiple Access

OHG Operators Harmonization Group

OMAP Operation, Maintenance and Administration Part

OMC Operation and Maintenance Center

OMSS Operation and Maintenance Subsystem

OS Operation System

OSI Open Systems Interconnection

P-IWMSC Packet Interworking MSC

PACCH Packet Associated Control Channel

PAD Packet Assembler/Disassembler

PAGCH Packet Access Grant Channel

PBCCH Packet Broadcast Control Channel

PBX Private Branch Exchange

PCCCH Packet Common Control Channel

PCH Paging Channel

PCN Personal Communication Network

PCS Personal Communication System

PDA Personal Digital Assistant

PDCH Packet Data Channel

PDN Public Data Network

PDN Packet Data Network

PDP Packet Data Protocol

PDTCH Packet Data Traffic Channel

PDU Protocol Data Unit

PLL Physical Link Layer

PLMN Public Land Mobile Network

PNCH Packet Notification Channel

PPCH Packet Paging Channel

PRACH Packet Random Access Channel

PSK Phase Shift Keying

PSPDN Packet Switched Public Data Network

PTCCH Packet Timing Advance Control Channel

PTM Point-to-Multipoint Service

PTM-G Point-to-Multipoint Service – Group Call

PTM-M Point-to-Multipoint Service – Multicast

P-TMSI Packet Temporary Mobile Subscriber Identity

PTP Point-to-Point Service

QN Quarter Bit Number

QoS Quality of Service

RA Rate Adaptation

RA Routing Area

RACH Random Access Channel

RAI Routing Area Identity

RAND Random Number (for authentication)

REVC Reverse Charging

RFCH Radio Frequency Channel

RFL Physical RF Layer

RFN Reduced TDMA Frame Number

RLC Radio Link Control layer

RLL Radio in the Local Loop

RLP Radio Link Protocol

ROSE Remote Operation Service Element

RPE Regular Pulse Excitation

RR Radio Resource Management

SACCH Slow Associated Control Channel

SAT SIM Application Toolkit

SATIG Satellite Interest Group

SB Synchronization Burst

SCCP Signaling Connection Control Part

SC-FDMA Single Carrier FDMA

SCH Synchronization Channel

SCN Sub Channel Number

SCP Service Control Point

SDCCH Stand-alone Dedicated Control Channel

SDMA Space Division Multiple Access

SFIR Spatial Filtering for Interference Reduction

SGSN Serving GPRS Support Node

SID Silence Descriptor

SIM Subscriber Identity Module

SM-CP Short Message Control Protocol

SM-RP Short Message Relay Protocol

SMLC Serving Mobile Location Center

SMS Short Message Service

SMS-GMSC Short Message Service – Gateway MSC

SMS-IWMSC Short Message Service – Interworking MSC

SMS-SC Short Message Service – Service Center

SMSCB Short Message Service Cell Broadcast

SMSS Switching and Management Subsystem

SN Subscriber Number

SNDCP Subnetwork Dependent Convergence Protocol

SNR Serial Number

SNR Signal to Noise Ratio

SOSS Support of Operator Specific Services

SP Signaling Point

SPC Signaling Point Code

SRES Session Key (for authentication)

SS Supplementary Services

SSL Secure Socket Layer

SSP Service Switching Point

TA Terminal Adaptor

TA Timing Advance

TAC Type Approval Code

TACS Total Access System

TAF Terminal Adaptation Function

TBF Temporary Block Flow

TCAP Transaction Capabilities Application Part

TCH Traffic Channel

TCP Transmission Control Protocol

TD-CDMA Time Division CDMA

TDD Time Division Duplex

TDMA Time Division Multiple Access

TD-SCDMA Time Division Synchronous CDMA

TETRA Terrestrial Trunked Radio

TFI Temporary Flow Identifier

TFO Tandem Free Operation

TID Tunnel Identifier

TLLI Temporary Logical Link Identifier

TLS Transport Layer Security

TMN Telecommunication Management Network

TMSI Temporary Mobile Subscriber Identity

TN Time Slot Number

TOA Time of Arrival

TSC Training Sequence Code

UDI Unrestricted Digital Information

UDP User Datagram Protocol

Um Air/Radio Interfacee

UMTS Universal Mobile Telecommunication System

UPT Universal Personal Telecommunication

URAN UMTS Radio Access Network

URL Universal Resource Locator

USF Uplink State Flag

UTRA UMTS Terrestrial Radio Access

UTRAN UMTS Terrestrial Radio Access Network

UUS User to User Signaling

UWCC Universal Wireless Communications Consortium

VAD Voice Activity Detection

VBS Voice Broadcast Service

VGCS Voice Group Call Service

VLR Visited Location Register, VLR Nummer

W3C World Wide Web Consortium

WAE Wireless Application Environment

WAP Wireless Application Protocol

WBMP Wireless Bitmap (Format)

W-CDMA Wideband CDMA

WDP Wireless Datagram Protocol

WLL Wireless Local Loop

WML Wireless Markup Language

WRC World Radio Conference

WSP Wireless Session Protocol

WTA Wireless Telephony Application

WTLS Wireless Transport Layer Security Protocol

WTP Wireless Transaction Protocol

WWW World Wide Web

XML Extensible Markup Language

XSL Extensible Style Language

References

Ambrosch, W. D., Maher, A. and Sassceer, B. (1989) *The Intelligent Network*, Springer, Berlin.

Begin, G. and Haccoun, D. (1994) Performance of sequential decoding of high-rate punctured convolutional codes. *IEEE Transactions on Communications*, **42**(2), 966–978.

Bertsekas, D. and Gallager, R. (1987) *Data Networks*, Prentice-Hall, Englewood Cliffs, NJ.

Bettstetter, C., Vögel, H. J. and Eberspächer, J. (1999) GSM Phase 2+ General Packet Radio Service GPRS: Architecture, protocols, and air interface. *IEEE Communications Survey (Special Issue on Packet Radio Networks)*.

Bluetooth SIG (2008) http://www.bluetooth.com.

Bocker, P. (1990) *ISDN – Das Diensteintegrierende digitale Nachrichtennetz*, 3rd edn, Springer, Berlin.

Bossert, M. (1991) D-netz-grundlagen – Funkübertragung in GSM-Systemen, Teil 1 und 2. *Funkschau*.

Bossert, M. (1999) *Channel Coding for Telecommunications*, John Wiley & Sons, Ltd, Chichester.

Brasche, G. and Walke, B. (1997) Concepts, services, and protocols of the new GSM Phase 2+ general packet radio service. *IEEE Communications Magazine*, **35**(8), 94–104.

Bruhn, D., Ekudden, E. and Hellwig, K. (2000) *Adaptive Multi-rate: A New Speech Service for GSM and Beyond*. Proceedings 3rd ITG Conference Source and Channel Coding (*ITG Technical Report 159*), VDE-Verlag, Munich.

Buckingham, S. (1999) Mobile positioning – an introduction http://www.mobilepositioning.com.

CCITT Recommendation M.3010. *Principles for a Communications Management* (1992).

CCITT Recommendation M.3020. *TMN Interface Specification Methodology* (1992).

Damosso, E. (ed.) (1999) Digital mobile radio towards future generation systems. *Final Report of COST Action 231*, European Commission.

David, K. and Benkner, T. (1996) *Digitale Mobilfunksysteme*, B. G. Teubner, Stuttgart.

Decker, P. and Pertz, U. (1993) Simulative Leistungsbewertung der nichttransparenten Fax-Übertragung im GSM-System. Walke, B. (ed.), *Informationstechnische Gesellschaft im VDE: Mobile Kommunikation, Lectures of the ITG-Fachtagung*, Neu-Ulm (*ITG Technical Report 124*), VDE-Verlag, Berlin.

Eberspächer, J., Vögel, H.-J. and Bettstetter, C. (2001) *GSM – Switching, Services and Protocols*, 2nd edn, John Wiley & Sons, Ltd., Chichester.

Eberspächer, J. (ed.) (1999) *Vertrauenswürdige Kommunikation*. Proceedings of the Münchner Kreis Hüthig, Heidelberg.

ETSI (2008) http://www.etsi.org.

Faccin, S., Hsu, L., Koodli, R., Le, K. and Purnadi, R. (1999) GPRS and IS-136 integration for flexible network and services evolution. *IEEE Personal Communications*.

Fuhrmann, W., Brass, V., Janßen, U., Kühl, F. and Roth, W. (1993) Digitale kommunikationsnetze. *Tutorium ITG/GI Fachtagung Kommunikation in Verteilten Systemen KiVS*, Munich.

Furuskär, A., Nälsund, J. and Olofsson, H. (1999) GPRS – enhanced data rates for GSM and TDMA/136 evolution. *Review*, Ericsson.

Geng, N. and Wiesbeck, W. (1998) *Planungsmethoden für die Mobilkommunikation*, 1st edn, Springer, Berlin.

Glitho, R. H. and Hayes, S. (1995) Telecommunications management network: vision vs. reality. *IEEE Communications Magazine*, **33**(3), 47–52.

Goodman, D. J. (2000) The wireless internet: promises and challenges. *IEEE Computer*, **33**(7), 36–41.

Gottschalk, H. (1993) Zeichengabetechnische Anbindung digitaler Mobilfunknetze an das Festnetz der Telekom. Walke, B. (ed.), *Informationstechnische Gesellschaft im VDE: Mobile Kommunikation, Lectures of the ITG-Fachtagung*, Neu-Ulm (*ITG Technical Report 124*), VDE-Verlag, Berlin.

Granbohm, H, and Wiklund, J. (1999) GPRS – general packet radio service. *Review*, Ercisson.

GSM World (2008) http://www.gsmworld.org.

Hagenauer, J., Seshadri, N. and Sundberg, C.-E. W. (1990) The performance of rate compatible punctured convolutional codes for digital mobile radio. *IEEE Transactions on Communications*, **38**(7), 966–980.

Hartmann, C. and Eberspächer, J. (2001) Adaptive radio resource management in F/TDMA cellular networks using smart antennas. *European Transactions on Telecommunications*, **12**(5), 439–452.

Hartmann, C. and Vögel, H. (1999) Teletraffic analysis of SDMA-systems with inhomogeneous MS location distribution and mobility. *Wireless Personal Communications*, **11**(1), 45–62.

Hata, M. (1980) Empirical formula for propagation loss in land mobile radio services. *IEEE Transactions on Vehicular Technology*, **29**(3), 317–325.

ISO/IEC 33091991. *Information Technology – Telecommunications and information exchange between systems – High-level data link control (HDLC) procedures – Frame structure*.

ITU-T Recommendation Q.735. *Multi-Level and Preemtion (MLPP)*.

ITU-T Recommendation V.110. *Support of Data Terminal Equipment (DTEs) with V-Series Type Interfaces by an Integrated Services Digital Network (ISDN)*.

Johannesson, R. and Zigangirov, K. S. (1999) *Fundamentals of Convolutional Codes*. IEEE Press, Piscataway, NJ.

Junius, M. and Marger, X. (1992) *Simulation of the GSM Handover and Power Control Based on Propagation Measurements in the German D1 Network*. Proceedings of the 5th Nordic Seminar on Digital Mobile Radio Communications (DMR V), Helsinki.

Kalden, R., Meirick, I. and Meyer, M. (2000) Wireless Internet access based on GPRS. *IEEE Personal Communications*, **7**(2), 8–18.

Kallel, S. (1995) Complementary punctured convolutional (CPC) codes and their applications. *IEEE Transactions on Communications*, **43**(6), 2005–2009.

Kleinrock, L. (1975) *Queueing Systems – Vol. 1: Theory*, John Wiley & Sons, Ltd., New York.

Laitinen, M. and Rantale, J. (1995) Integration of intelligent network services into future GSM networks. *IEEE Communications Magazine*, **33**(6), 76–86.

Lee, W. C. Y. (1989) *Mobile Cellular Telecommunication Systems*, McGraw-Hill, New York.

Lin, Y. B. (1997) OA&M for the GSM network. *IEEE Network*, **11**(2), 46–51.

Mende, W. (1991) *Bewertung ausgewählter Leistungsmerkmale von zellularen Mobilfunksystemen*, PhD Thesis, Fernuniversität, Hagen.

Molkdar, D. and Featherstone, W. (2001) *System Level Performance Evaluation of EDGE Compact*. Proceedings of the IEEE Symposium on Personal Indoor and Mobile Radio Communication (PIMRC 2001), San Diego, CA.

Molkdar, D., Featherstone, W. and Lambotharan, S. (2002) An overview of EGPRS: the packet data component of EDGE. *Electronic and Communication Engineering Journal*, **14**(1), 21–38.

Mouly, M. and Pautet, M. B. (1995) Current evolution of the GSM systems. *IEEE Personal Communications*, **2**(5), 9–19.

Natwig, E. (1998) Evaluation of six medium bitrate coders for the pan-European digital mobile radio system. *IEEE Journal on Selected Areas in Communications*, **6**(2), 324–331.

Okumura, Y. (1968) Field strength and its variability in VHF and UHF land-mobile radio-services. *Review of the Electrical Communications Laboratory*, September/October, pp. 825–873.

Proakis, J. G. (1995) *Digital Communications*, 3rd edn, McGraw-Hill, New York.

Rappaport, T. S. (2002) *Wireless Communications: Principles and Practice*, Prentice-Hall, Englewood Cliffs, NJ.

Sahasrabuddhe, L. H. and Mukherjee, B. (2000) Multicast routing algorithms and protocols: a tutorial. *IEEE Network*, **14**(1), 90–102.

Sahin, V. (1993) Telecommunications Management Network – principles, models, an applications. *Telecommunications Network Management into the 21st Century*. IEEE Press, Piscataway, NJ.

Schmidt, S. (1993) Management (Operation & Maintenance) von GSM Base Station Subsystemen. Walke, B. (ed.), *Informationstechnische Gesellschaft im VDE: Mobile Kommunikation, Lectures of the ITG-Fachtagung*, Neu-Ulm (*ITG Technical Report 124*), VDE-Verlag, Berlin.

Schwartz, M. (2005) *Mobile Wireless Communications*, Cambridge University Press, Cambridge, UK.

Sexton, T. A. (2000) *EDGE Compact Control/Data Bandwidth Dimensioning*. Vehicular Technology Conference (VTC 2000).

Steele, R. (1992) *Mobile Radio Communications*, Pentech Press, London.

Steele, R. and Hanzo, L. (1999) *Mobile Radio Communications*, 2nd edn, John Wiley & Sons, Ltd., Chichester.

Tanenbaum, A. S. (1996) *Computer Networks*, 3rd edn, Prentice-Hall, Englewood Cliffs, NJ.

Towle, T. S. (1995) TMN as applied to the GSM network. *IEEE Communications Magazine*, **33**(3), 68–73.

Tran-Gia, P. (1996) *Analytische Leistungsbewertung verteilter Systeme*, Springer, Berlin.

Vögel, H. J., Johr, H. and Grom, A. (1995) Messung und verbesserte Markov-Modellierung transparenter GSM-Datendienste. Walke, B. (ed.), *Mobile Kommunikation, Lectures of the ITG-Fachtagung*, Neu-Ulm (*ITG Technical Report 135*), VDE-Verlag, Berlin.

Walke, B. (1999) *Mobile Radio Networks: Networking and Protocols*, John Wiley & Sons, Ltd., Chichester.

WAP Forum (1999a) http://www.wapforum.org.

WAP Forum (1999b) *Official Wireless Application Protocol – The Complete Standard*. John Wiley & Sons, Ltd, Chichester.

WAP Forum (1999c) *Wireless Application Protocol, White Paper*, WAP Forum.

Watson, C. (1993) Radio Equipment for GSM. *Cellular Radio Systems*, Artech House, Norwood, MA.

WWW Consortium (2008) http://www.w3.org.

Xu, G. and Li, S. Q. (1994) Throughput multiplication of wireless LANs for multimedia services: SDMA protocol design. *IEEE Globecom 94*, San Francisco, CA.

Index